SKY TIGER

By the same author:

Double Mission:
  Fighter Ace and SOE Agent, Manfred Czernin DSO MC DFC

Fighter Leader:
  The Story of Wing Commander Ian Gleed DSO DFC

The Greatest Air Battle

Wings of Freedom:
  Twelve Battle of Britain Pilots

Sky Tiger

The Story of
Group Captain Sailor Malan, DSO DFC

NORMAN L. R. FRANKS

WILLIAM KIMBER · LONDON

First published in 1980 by
WILLIAM KIMBER & CO. LIMITED
Godolphin House, 22a Queen Anne's Gate,
London, SW1H 9AE

ISBN 0 7183 0487 X

Typeset by Print Co-ordination
and printed and bound in Great Britain by
Redwood Burn Limited, Trowbridge and Esher

Contents

List of Illustrations

Maps and Diagrams

Acknowledgements

I wish to thank the following people for their help and encouragement during the research and writing of this book. I am more than grateful for their time and kindness, not to mention patience.

Air Chief Marshal Sir Harry Broadhurst GCB KBE DSO DFC AFC, Air Vice-Marshal H.A.C. Bird-Wilson CBE DSO DFC AFC, Air Vice Marshal J.E. Johnson CB CBE DSO DFC, Air Commodore A.C. Deere DSO OBE DFC, Air Commodore J.A. Leathart CB DSP, Air Commodore D.M. Strong CB AFC, Group Captain M. Duke-Woolley DSO DFC, Group Captain D.P.D.G. Kelly DFC, Wing Commander R.J.E. Boulding, Wing Commander N.F. Duke DSO OBE DFC AFC, Wing Commander H.M. Stephen DSO DFC, Wing Commander R.R.S. Tuck DSO DFC, Squadron Leader W.A.K. Igoe, Squadron Leader H. Szczesny VM DFC KW.

To my friend Chaz Bowyer for help with original research and photographs, Martyn Ford-Jones, friend and fellow enthusiast for his maps and who with his wife Val carried out some research for me, Mike Schoeman, and the ever helpful Ted Hine of the Imperial War Museum. To the staff of the Public Records Office, SAAF Pretoria and the Polish Institute and Sikorski Museum. As always to Amy Howlett and everyone at Wm Kimbers for their friendly help and encouragement, and of course to my wife and sons who have to put up with weeks of silence, followed by weeks of typing noises.

N.F.

CHAPTER ONE

Sailor

Adolph Gysbert Malan was the eldest son of William – known as Willie to all his friends – and Evelyn Malan. He was born in Wellington, South Africa, on the third day of October 1910.

The Malans are able to trace their family tree from its French Huguenot ancestry, to Jacques Malan, the first of the line to settle in South Africa as far back as 1694. Many families of the persecuted Huguenot religious body were forced to flee from their native France in the late 1600's, during the reign of King Louis XIV. The Huguenots, formerly a political party, had by that time, ceased its political background, being then only a religious group. They emigrated to various parts of the world, to Germany, to the Netherlands, some to England and others to South Africa. Jacques Malan started a farm in the open land which later became Wellington, his brother moved further west towards the Riebeeck Castle Mountain range.

Both lived side by side with Dutch settlers, both the Huguenots and the people from Holland growing together in and around Cape Province. The site of Wellington is in the valley of Charron (Val du Charron) which became known as Wagenmarkersvallei, the valley of the Wagonmakers. The native African Bantu tribes looked upon their new white neighbours with interest and with curiosity.

The Malans flourished; the families increased and re-settled in various parts of the Province over the years. When the Boer War began in 1899 there were Malans with sympathies for both sides, the Boers and the British. Willie Malan's sympathies were for the British but he was not actively engaged either way. The war raged about the countryside and a large raiding party under the leadership of General Jan Smuts began to attack and harass British patrols and outposts in the area of the Malan family home. Somehow, during a clash between Smuts' forces and the British at

Twenty-Four Rivers at Halfmanshof, near Pretoria, Willie Malan was shot from his horse, hit by two bullets. The first penetrated both thighs, the bone of the right leg being exposed. The second bullet ripped open his right arm. For days he lay between life and death with only an even chance of his shattered right leg not having to be amputated, but he gradually recovered, although he had to use crutches. In fact it was not until 1908, six years after the end of the Boer War, that Willie Malan finally threw away his crutches in disgust. With one leg two inches shorter than the other, still in considerable pain, he returned to his horses and his farming. He had a tough constitution – a toughness that was to show in his son Adolph.

In 1909 Willie married Evelyn Forde Jordan, the daughter, and one of seven children, of Alfred and Frances Jordan. The Jordans were English, from Leicester. In 1889 they had decided to emigrate to Wellington, New Zealand, stopping en-route in Cape Town. Hearing quite by chance that only 40 miles away was a town also called Wellington, Alfred Jordan went to see the area, liked it and decided they had come far enough. Evelyn was then two and a half years old.

After their marriage, Willie and Evelyn Malan lived in a farmhouse, Groenfontein, built by his grandfather, among orchards of plums and apricots with a stream passing the rear of the house, fed by the melting snows from the mountains. They remained at Groenfontein only a few years, during which time Adolph, and then a second son, Ralph, was born. They then moved to Klipvlei, just to the north-east of Wellington. Here Adolph and Ralph had the run of the farm and countryside. In 1915 they were on the move again, to Slent farm on the slopes of the Paardeberg, sixteen miles south of Wellington. Twenty miles further south stood the spectacular grandeur of Table Mountain.

It was at Slent that the young Malan was given his first gun by his father. It was not to be an auspicious start for the man whose shooting ability was to make him a legend in future years. It was a double barrelled shotgun and Adolph eagerly went outside to use it. Firing both barrels, and bruising his shoulder in the process, he managed to jam the two new cartridges in the gun's breech. When he ran back into the house to his mother for help, she was suddenly confronted by her young son carelessly aiming the gun in her direction. Angrily she took the weapon from him and it proved a long time before he was allowed to handle a firearm again.

Initially Adolph attended a local farm school, used by children from neighbouring farms, then went on to the school at Stellenbosch. This school was segregated into English-speaking and Afrikaan-speaking children. Adolph and his family spoke English. As his mother was from England this was natural, and Willie Malan and his wife had agreed that it was far better for them all to speak one language well rather than jumping constantly, as many Cape people did, from English to Afrikaan, often all in one sentence.

Unhappily for Willie Malan, the farm Slent did not prove a success. It was large, difficult to handle even for a man who was not crippled. Following the end of World War One and the depression that affected many parts of the Western World, Willie was forced to sell Slent at a considerably lower price than it was actually worth. The Malans moved back to Stellenbosch, then back to Wellington itself. The Malan family now numbered five, a third son, Francis, having been born. To add to their problems, Willie Malan suffered a breakdown at the time Slent was sold.

Back in Wellington, living in Malherbe Street, Adolph went to the Boys' High School while his mother helped to make ends meet by taking in boarders and turning her hand to dressmaking. There was also some help from her side of the family, the Jordans living only a short distance away.

At the end of 1923, with Adolph now 13½ years old, there came a radical change in his young life, when he suddenly told his parents that he wished to become a sailor. The first step in his plan was to attend as a cadet, the training ship *General Botha* (formerly HMS *Thames*). The *General Botha* was permanently anchored at Simonstown. The whole idea was something quite foreign to everyone and even the young Adolph had never even seen a ship in his life.

The *General Botha* had only been established for a few years, set up as a training ship for South Africans. It is not known exactly why Adolph should choose the sea as a future career but choose it he did. Perhaps it was all part of his make-up, for already it was apparent that the young man, who would one day gain undying fame in the skies of Europe, had certain traits. He could, if he so wished, be very stubborn, and he had developed a certain independence, with more than a little help from his father, who had firmly believed in encouraging self-reliance in all his children.

Adolph Malan joined the *General Botha* along with 60 new recruits, on 15 February 1924. He was one of the smallest cadets, and, with a smiling, cherubic face, quickly earned the nickname of 'Angel Face'. Life as a cadet proved far from easy but life generally was good, if hard. However, it was the good as well as the hard that helped to shape the young Malan boy. As the months grew into years he became anxious to go to sea. In the classroom he was only average; mathematics with its associated trigonometry and logarithms were for him a constant stumbling block. He eventually only received an ordinary pass, although in seamanship he achieved a First Extra certificate.

Then finally in 1927, he left the *General Botha*, travelled to Port Elizabeth to join the *Sandown Castle*, a ship of the Union Castle Steamship Line. As Adolph waved a farewell to his mother, the *Sandown Castle* steamed leisurely away from the quayside heading out into the Atlantic before setting course for New York. Quite suddenly the world was opening up for him.

Up to this time, Adolph Malan had given little, if any, thought to flying or aviation in general. Yet that same year of 1927 when he was making his first great sea adventure by crossing the mighty Atlantic Ocean, history was recording that one Charles Lindbergh was also making a similar journey, but by air, in the other direction, and alone.

Malan made no conscious acknowledgement of this famous feat of personal endurance and courage, but on his second voyage to New York, he saw the typical ticker-tape welcome given to Lindbergh on his return to America. Not unnaturally to the seventeen year-old Malan, and to many others, Lindbergh quickly assumed the status of boyhood hero and perhaps it was here that the first seeds were sown which would eventually lead the young South African to the air.

Yet that dream, if dream it was, was still a long way off. For the next eight years Adolph Malan continued to serve with the Union Castle Line although he was conscious that the sea as a lifelong career was not for him. But exactly what to do with his life was still shrouded in mystery. Life as a sailor was an uncertain one with no permanent roots. With a homely family background he was constantly looking for something, for somewhere, where he could put down these roots.

In his years at sea he remained a good and faithful servant to the Castle Line, serving in a total of nine different ships, trading generally to New York or Philadelphia, but sometimes to Hamburg in Germany or to the Continent of Europe. He saw his family occasionally – there were now four brothers at home, Ralph, Francis, Stanley and young Peter. They could see the changes in him, for his seafaring life had broadened his shoulders, tanned his face even more, and toned up his muscles. He had passed his Second Mate's Certificate and to add some spice to his life he spent time between voyages training with the Royal Naval Reserve with the rank of Sub-Lieutenant. In trying to decide on a real future, the RNR held some possibilities for him but it was still not a permanent home. There was an alternative, however, for several of his friends in the Union Castle Line had applied and been accepted for short service commissions with the Royal Air Force.

After a good deal of soul-searching he sat down and wrote to the RAF and was accepted. He was lucky (and so was the RAF as it turned out) for he had almost reached the age limit for application to a short service commission. His final sea trip was aboard HMS *Malaya* during RNR training manoeuvres – then it was a change of uniform from the dark blue of the Navy to the azure blue of the Royal Air Force.

At nearly twenty-five years of age, Adolph Malan had to start over again from the bottom, alongside younger men from both Britain and the Empire. However, one thing he did have was experience of life; and his knowledge of navigation and experience of actually having been an officer were definitely in his favour.

Basic ground training in any service can be a severe jolt from civilian life but to Malan, who had seen it all before, it was a piece of cake after the *General Botha* days. He quickly progressed to the more interesting flying stage of his RAF training. For this he went to Filton, near Bristol, where he was first shown how to fly. One of his companions at this time was David Strong:

'I first met Sailor Malan early in January 1936 when we both arrived at Filton airfield near Bristol to learn to fly at the Bristol Aeroplane Company Flying School. Officially we were known as Pupil Pilots and on successful completion of the course at Filton, which was scheduled to take seven weeks, we would be appointed acting pilot officers and go on for advanced pilot training at No 3 Flying Training School, Grantham. On the course at Filton there were ten pupil pilots and three pupil sergeants. These had been selected from the ranks of the RAF as suitable material to become sergeant pilots. Competition was very keen among the ranks of the service for nomination to be trained as sergeant pilots, selection was deemed a great honour.

'The pupil pilots were accommodated and boarded in a wooden hut converted into bedrooms by the local garage owner. Sailor and I shared a room which had two single beds but hardly enough space to swing a cat. At that time Sailor was two or three years older than the rest of us; his training and experience as a Deck Officer with the Union Castle Line and as an officer in the Royal Naval Reserve had made him justifiably confident in his ability to manage men and to deal with new situations.

'We were taught to fly on the De Havilland 82, better known as the Tiger Moth; the course required each trainee pilot to complete 25 hours dual instruction and 25 hours solo flying. Sailor was one of the first of us to fly solo and was rated as above average on completion. After the course at Filton we spent two weeks at the RAF Depot Uxbridge. Here we were joined by other officers and sergeant pilot trainees who had learned to fly at Hatfield and Hanwell, making up the course to 32 which was scheduled to become No 17 Course at No 3 FTS, Grantham, Lincs. The purpose of the two weeks at Uxbridge was to allow the officer pupils to visit London tailors to obtain uniforms and also to instruct us in drill and in Mess Rules and etiquette. Sailor's previous experience once again stood out; he had done most of it before.

'We were all 'confined to camp' during our stay at Uxbridge

apart from our visits to the tailor of our choice in London. On the first Sunday morning after church parade, however, Sailor just walked away to the station and left to see his girl friend in London, thus displaying early in his RAF career, a quality of initiative which to the narrow minded disciplinarian might appear more like breaking the rules. In the event, his absence went unnoticed.

'On arrival at No 3 FTS, Sailor was appointed the Course Commander and remained so throughout. Two pilot officers joined the Course at Grantham. Both were university entrants and were therefore Gazetted "pilot officers" while the rest of us being short service officers, were Gazetted "acting pilot officers". No 17 Course had some distinguished members. One of the university entrants, Paul Holder, became Air Marshal Sir Paul Holder (KBE CB DSO DFC FRAeS); two New Zealanders later became Chiefs of Air Staff RNZAF, Air Vice Marshal I.G. Morrison, who was followed by Air Vice Marshal C.A. Turner.

'At Grantham we flew Hawker Hinds and Audaxes and completed 100 hours during the course of nine months. In the second term of the course, allocations were made to what were known as the Fighter Flight or the Bomber Flight and future training was directed towards fighter or bomber squadrons as appropriate. Sailor had always been determined to become a fighter pilot and, needless to say, he did.

'Sailor was not an accomplished games player – the years at sea had not given him much chance to become so, but he was a keen member of the pack in the 3 FTS Rugby XV and played with much spirit. Socially he was "one of the chaps" despite a natural reserve. He did his share of drinking and was a good companion on party occasions. Surprisingly, he could play a pretty tune on the mough-organ as we discovered in a pub after a rugger match.

Altogether, I spent just over eleven months of 1936 in close proximity to Sailor Malan. I admired him and liked him.'

At No 3 FTS the order of the day was just to learn to fly. There were two courses, each lasting nine months and overlapping by six months. One young man who had joined the course previous to Malan's was Robert Stanford Tuck, who also recalls Malan at Grantham.

> He was a very good looking, solid, square shouldered, blond chap. We had a certain kinship, both having been cadets in the Merchant Service. We were kept pretty busy, for it was nearly a year's intensive training with ground studies, meteorology, air navigation, rigging of aircraft, engines, and of course a heavy flying programme. Lots of sport – Sailor played a lot of rugger, while I was more for swimming and fencing. It was a pretty full life.

Like Malan, Tuck was also destined to become a household name within a few short years – one of Britain's air heroes. He had been a cadet with the Lamport and Holt Shipping Line before he too decided to seek his future in the air in 1935. Although he was several years younger than Malan, they were to become firm friends in the years ahead. And their joint, if separate, decisions to leave the sea for an air force career was to cost a future German Luftwaffe dearly.

At Grantham, Adolph, with his seafaring background, gained yet another nickname – 'The Admiral'. At least it was better than 'Angel Face'.

Finally at Grantham Malan flew the Gloster Gauntlet. This was the last of the open cockpit fighters to serve with the RAF and was a forerunner of the famous Gladiator. Yet it was to be in a Hawker Fury that Malan was to have one of his first moments of anguish in the air. He had just taken-off and was only about 100 feet off the ground when the Fury's engine cut out. The aeroplane dipped forward, only seconds away from disaster. Malan quickly steepened the angle of glide and luckily the engine picked up, the South African pulling away just inches from the ground.

Soon, however, his training was completed and then came the eagerly awaited postings to operational squadrons. Finally the names were posted on the bulletin board. Acting Pilot Officer A.G. Malan posted to Number 74 (Fighter) Squadron, at RAF Hornchurch.

CHAPTER TWO

Fighter Pilot

With his posting to Number 74 (Fighter) Squadron, Malan's dreams of becoming a fighter pilot were realised. The official posting date was 20 November 1936, and he arrived with two colleagues he had teamed up with at training school, Paddy Treacy and Paddy Byrne.

Number 74 Squadron, the famed 'Tiger', had a proud history already, dating from World War One. In that conflict it had flown single-seat SE5 fighters, or scouts as they were more commonly known in those days, during the last year of the war – 1918. In France, with such leaders as Mick Mannock and Keith Caldwell, it had more than proved itself on the Western Front. Among its redoubtable air fighters was one Ira 'Taffy' Jones, who had claimed 40 victories. In World War Two Taffy Jones took a keen interest in the boys of his former squadron and often visited the new generation of Tigers. He had known the great Mick Mannock and was to liken him to Sailor Malan in many ways.

When Malan arrived on 74 Squadron, its Commanding Officer was Squadron Leader D.S. 'Brookie' Brookes who had previously been one of its flight commanders. The squadron had only recently returned to England from Malta and was equipped with Hawker Demon two-seat fighters. It was based at RAF Hornchurch in Essex where it joined 54 and 65 Squadrons who flew Gauntlets. In 65 Squadron was Malan's friend from Grantham, Robert Stanford Tuck. They saw each other most days, would drink together in the officers mess and often played squash. Tuck recalls:

> I recall great arguments and indeed financial jugglings we used to have when his car would be out of action – a very old, big, Austin – and I would hire him my equally old Vauxhall for a very nominal sum. The cash enabled me to buy enough beer in

the evening while he rushed over to see his dear Lynda at Ruislip. The money would be adjusted when my car was out of action and I would have to hire his for the night.

Lynda was Lynda Irene Fraser who lived with her parents in Ruislip, in Middlesex. They had first met in 1930 when with some naval friends he had been invited to her parents' home. Lynda was then sixteen years old. Now their relationship had flourished into something more than just friendship.

For several months 74 Squadron was kept busy, training its many new intake of pilots, then in April 1937 came a change of aeroplane. The Demons went out, the Gloster Gauntlet came in. It was at that time that the now famous Tiger's head insignia first appeared on the fin of the silver-coloured peace-time aircraft.

In July 1937 Pilot Officer W.E.G. 'Tink' Measures arrived on the squadron, posted in from the RAF College at Cranwell.

The squadron's Annual Armament Camp that year took place between 4 and 24 October, at Woodford in Dorset (which later became Warmwell, and famous as an advanced fighter airfield in WW2). By far the most outstanding pilot in air gunnery in the squadron was Malan. His was a natural gift, especially in deflection shooting, the art of judging during a turn, where your bullets will be ahead of the target, ie: enemy aircraft, in order for the target to fly into them. It was always important for a fighter pilot to be able to shoot well, although it was not always as easy as Malan made it seem. There were a number of fighter pilots in the war who were notoriously bad at air gunnery but many of them used their superb flying skills to get in close – so close they couldn't miss! With Malan's gift at shooting and his first-class flying skill he would be head and shoulders above many of his contemporaries when the war came.

Less than a year after joining 74 Squadron, Malan, who had very quickly been demoted from 'The Admiral' to just plain 'Sailor', and Paddy Treacy were made acting flight commanders – recognition of their flying abilities and leadership.

Initially, both Treacy and Byrne, both Irishmen of course, had been put in the same flight, but as one came from Southern

Hawker Demon of 74 Squadron, flown by Flying Officer A G Malan, 1937.

K5337, 74 Squadron's first Gloster Gauntlet, 1937.

Tiger's head on the tail of the CO's Gauntlet – S/Ldr D S Brookes.

Ireland, the other from the North, they fought like hell! It rather reminds this author of the story told to a group of aviation enthusiasts by Marshal of the Royal Air Force Sir Arthur 'Bomber' Harris. Apparently one of his bomber crews in WW2 had two gunners, again one from the South and one from the North of Ireland. On one mission whilst on the way to the target, an argument developed between them, centering on the Eire Prime Minister Eamon De Valera who maintained a strict policy of neutrality for the Republic of Ireland. As the target came up, their pilot told them both to shut up and concentrate on watching for night fighters. With a final retort, the man from the south, settling himself behind his guns, yelled at his northern companion, 'Well, at least De Valera kept us out of the war!' Harris went on to say that during the war some of his best fighting men came from Southern Ireland, usually on the basis of, 'Is this war a private fight or can anyone join in?'

On 27 November 1937 Acting Pilot Officer Ronald George Temple-Harris joined the squadron from Number 10 FTS, Ternhill.

As 1938 progressed so too did 74 Squadron. It was to be a momentous year in many respects and for Sailor Malan especially so. It began on 2 April, the day after the RAF's twentieth birthday, when he and Lynda Fraser were married at St Martin's Church, Ruislip.

The Squadron acquired a new CO towards the end of April when Squadron Leader George Edward Sampson arrived from Malta to take over from Squadron Leader Brookes, who had left on the 10th. Sampson quickly took over the reins of command, and Malan and Treacy, his two stalwart flight commanders, continued to build the squadron into an even more efficient unit. The political scene in Europe gave everyone cause for concern and the RAF, in terms of strength and equipment, were far from ready should a world conflict be forced upon them at this moment.

In May Malan led his A Flight to Gravesend for an Empire Air Day display. This was always good for a boost of morale for 'Joe Public', showing him how strong and wonderful his air force was! Something akin to the old adage, 'If you can't fight, wear a big hat!' B Flight, meantime, flew a display at Hornchurch, Paddy Treacy giving a splendid solo aerobatic display at both Woodley and Hanworth.

The squadron's diary gives us a picture of the 1938 scene: 10 June, – General Mobilization Exercise, lasting one week. 1st to 8 August – Searchlight Co-operation and Home Defence Exercises. Then on 19 September came the Annual Armament Training Camp at Sutton Bridge. Unfortunately it coincided with the Munich Crisis and 74 had to return to Hornchurch, where a readiness state was maintained. Their Gauntlet fighters were hastily repainted from their colourful peacetime silver to a camouflaged green and brown. The crisis passed and the RAF, Britain and the world breathed again. It was, however, a shot in the arm for everyone. Preparations for the now strong possibility of war, began.

A new pilot arrived at Hornchurch at the end of the summer; Pilot Officer Alan Christopher Deere, a New Zealander, came to join 54 Squadron. The peacetime air force used to have block leave periods in those days, and 54 Squadron was on leave when Deere arrived. He was, therefore, attached to Malan's A Flight in 74 Squadron for a fortnight or so. Deere became firm friends with the South African and would fly with him in the mid-war years.

One of Deere's memories of Malan at this early period was during a rugger match. Malan played as a forward, Paddy Byrne was usually a half-back. Byrne's usual tactic if he got the ball was to grip it firmly, then with head well down he would charge into the opposition, or whoever else was in his way. During one match he did just this but Malan, who was of course on Byrne's side, failed to get out of the way quickly enough. Byrne hit Malan squarely in a very sensitive area mid-way between the stomach and the knees! Malan doubled up and rolling on the ground was heard to scream out, 'Byrne, you little twerp, are you trying to finish me?'

On 29 October, Acting Pilot Officer John Connell Freeborn joined the squadron from Number 8 FTS. He was an A Class Reservist, the first to join the Tigers.

Just over a week later 74 Squadron succeeded in winning the preliminary round of the Flight Fighter Attack Competition in the

Sir Philip Sassoon Challenge Trophy. It was held at Northolt where they beat their rival units from Hornchurch, 54 and 65 Squadrons. And this was despite the fact that 74 was still flying its old Gauntlets, while 54 and 65 had progressed to the Gloster Gladiator. The 74 squad was led by Sailor Malan, his five companions being Flying Officer Tom Rowland, Pilot Officers S.T. 'Charlie' Meares, Haywood and Sergeants Ian Hawken and 'Polly' Flinders. This was no new team just for the competition; they had been flying together for almost two years.

Qualifying for the final, which took place on 25 November, the team pulled off a spectacular victory. Alan Deere recalls watching the competition from the ground, as the judges had obviously to be able to see the pilot's flying skill. Deere was impressed with Malan's precision and it was obvious to him that Malan and his A Flight were far and above all the other flight finalists. Their crisp and practised quarter attacks (simulated of course) were superb. The Flight Fighter Attack Trophy was theirs.

The Sassoon Trophy had various categories, and 74 Squadron just failed to qualify in the Map Reading Competition, obtaining 94 out of a possible 100 points.

Then in the second week of December, the long promised and eagerly awaited change of aeroplane took a step nearer fruition when Malan and Treacy were attached to RAF Duxford for a course of instruction on Spitfires. Duxford was the home of 19 Squadron, which had the distinction of being the first RAF squadron to equip with the beautiful, acquiline, Supermarine Spitfire. This course was repeated between 11 and 19 January 1939, again with 19 Squadron. Malan's friend Bob Stanford Tuck was another pilot to be attached to Duxford, for 65 Squadron too were soon to change over to Spitfires. 54's change to Spitfires soon followed.

Re-equipment for the Tigers commenced on 13 February. On that day Malan and Treacy went to Eastchurch to collect the first Spitfires, and the two flight commanders returned in company with a third Spitfire, flown on this special occasion by 74's ex-boss Brookie Brookes. The squadron also had a Fairey Battle bomber delivered for the senior pilots to instruct other pilots on the new monoplane type with its enclosed cockpit, retractable undercarriage and Merlin engine. Then on the 16th, 74 at last said farewell to its old Gauntlets as they were flown off to Number 24 Maintenance Unit. By early March the transition to the Spitfire was complete. 74 Squadron was destined to fly various marks of the Spitfire to the end of World War Two with just one short period on Hurricanes in 1943.

On 1 March Pilot Officer P.W.O. 'Boy' Mould arrived on the squadron, in from RAF Cranwell. He only remained with the Tigers until the end of April when he went to 1 Squadron. With them, which became part of the Advanced Air Striking Force in France when the war began, he claimed the first RAF fighter victory in France, a Dornier 17 bomber, which he shot down on 30 October 1939.

Amongst another batch of new pilots to arrive in April was Pilot Officer Harbourne Mackay Stephen. 'Steve' was to remain with 74 until early 1941 and became one of its most successful air fighters. Stephen remembers being frightened to death when he first came into contact with Sailor Malan.

> He was so completely professional, so on the ball. Already he had quite a name for himself at Hornchurch as being probably the best pilot on the station, and was rapidly becoming well known within Fighter Command as a tremendous shot. We used to do our gunnery practice at Bradwell and Sailor's results were always well ahead of those of anyone else.

Stephen also remembers Malan's strict discipline, a discipline obviously inherited from his early Navy days. He was very forthright in his views and in the way he handled his flight and later the squadron. Strict discipline in the air was essential when flying with Malan; keeping his men tucked in close he would yell over the radio, 'Get your machine tucked right in and don't bloody well move away!'

The squadron pilots soon became proficient on the new aeroplanes but Malan, Treacy and the CO were all kept particularly busy building the squadron through an exacting and detailed training programme. Towards the end of April, Malan's A Flight flew to Upper Heyford for four days, taking part in affiliation exercises with 18 and 57 Bomber Squadrons.

There was, of course, keen rivalry between the personnel of all three squadrons at Hornchurch. It was good for morale as well as giving an excitement, an edge, to all their many activities both in the air and on the ground. These men were, after all, the professionals, the regulars. When the war began in September 1939 it would be these same professionals who would bear the brunt of the air war until rapid expansion brought new blood into the service. There were, nevertheless, a few Volunteer Reservists (VRs) beginning to arrive as 1939 progressed. 74 Squadron welcomed Sergeant William Malcolm Skinner RAFVR on 10 June, in from Number 5 E & R FTS (Elementary and Reserve Flying Training School) Hanworth, to receive his advanced training with a front line service squadron. This was part of the overall training scheme in the late 1930's, whereby a pilot would spend a period with a civilian school then go onto service types with the RAF to complete his training.

Exactly one month after Skinner's arrival, 74 Squadron flew to Le Bourget aerodrome, near Paris, to take part in the Bastille Day Celebrations in conjunction with units of the French Air Force.

On 5 August Acting Pilot Officer Richard Denis 'Bertie' Aubert arrived from Number 6 FTS at Netheravon. He was followed by Pilot Officer Derek Hugh Dowding, son of Air Chief Marshal Hugh Dowding, Commander in Chief of Fighter Command. He arrived from RAF Cranwell.

Then quite suddenly the halcyon days of peace were at an end. On 1 September, Germany invaded Poland. Two days later, one hour before mid-day, Great Britain was once again at war with Germany.

Tiger Squadron was on the alert from the very beginning. The third day of September was a strange day, an unreal day. What everyone on the squadron had been busily training for, half expecting, yet hoping would not come was now upon them.

Sailor Malan flew his first 'war' mission shortly after midnight – the 4th. Leading the squadron's Red Section, with Tink Measures and Sergeant Hawken, Malan, flying Spitfire K9864, was airborne at 2.50 am, following a report that enemy night bombers were flying over Surrey. Paddy Byrne led off Yellow Section into the darkness shortly afterwards. The 'enemy bombers' proved to be RAF aeroplanes – and Malan landed back at 4.05.

At this time it was believed that any assault by the German Luftwaffe would most likely come from the east, direct from Germany, and probably at night, in order to avoid flying over hostile (France and Belgium) territory in daylight. It is strange, therefore, that Britain's night defences were so poor and that day fighters were alerted and ordered up into the night sky over England. Only when France, Belgium and Holland fell did the threat of attack by air come from the south and south-east; and with these territories occupied the Luftwaffe could come in daylight.

However, the 4th did bring one thing more substantial. Pilot Officer John Colin Mungo-Park arrived from the RNAS Ford where he had been with the Fleet Air Arm. He was destined to lead and command the Tigers in 1941.

With everyone being so excited and keyed-up now that the war had finally started it was almost inevitable that mistakes in aircraft identification would occur sooner or later. It was to happen often during the war, but the first instance took place on 6 September, 74 Squadron being involved.

Twelve Spitfires, six from each flight, were scrambled at 6.45 am, Malan leading in K9932. A raid had been reported and the pilots were fully expecting to encounter hostile aeroplanes. So, when shortly afterwards they spotted what they thought were Messerschmitt fighters, two pilots, Paddy Byrne and Johnny Freeborn, dived to the attack. However, the Messerschmitts were in fact Hawker Hurricanes of 56 Squadron. Two were shot down, Pilot Officer Halton-Harrop being killed, Pilot Officer Rose luckily surviving. Both Byrne and Freeborn had to attend a Court Martial the following month and both were acquitted.

During November the squadron took its turn to operate from Hornchurch's satellite airfield at Rochford. They went through a period of minor accidents but then on the 20th came the squadron's first combat victory of the war.

Flying Officer Measures (K9932), Pilot Officer Temple-Harris

(K9864) and Sergeant Flinders (K9870) were sent off at 12.05 pm to intercept a Heinkel 111 flying at 27,000 feet, fifteen miles east of Southend, the German flying on an easterly course. They found the bomber and all three pilots attacked as it dived towards the safety of cloud cover. In fact the Heinkel had been seen over Hornchurch at 12.17 pm. Following the attacks, by which time the Heinkel had reduced height considerably, the bomber began to leave a long trail of smoke. It was then lost in the cloud but the following day two wounded survivors were rescued from the sea. 74 had opened its account with the Luftwaffe.

Over the severe winter of 1939-40, 74 flew when it could, Malan and Treacy ever mindful of the urgent need to bring every pilot under them up to the most efficient state possible. Only practice and time in the air was any good now, with the ultimate test of actual combat to see if they had learnt their lessons well.

On 25 February 1940 Squadron Leader Francis Lawrence White arrived as a supernumerary, pending his appointment of CO of the Tigers on 1 March. He came from a training post at RAF Cranwell.

The 'phoney war' – the period so called because so little happened (at least in the West), lasted until early May 1940. The squadrons in France had occasional clashes with the Luftwaffe and even over Britain there was the odd reconnaissance aeroplanes to chase or engage. Otherwise nothing. But it was all about to happen. The professionals were about to test all their months or years of training. The real war was about to start, a war in which 74 Squadron's A Flight commander, Sailor Malan, would shine.

Three days before the 'balloon went up' in France, on 10 May 1940, Pilot Officer Peter Cape Beauchamp St John joined the Tigers from 501 Squadron.

On the 10th, all leave was suddenly cancelled. The Germans had finally started the war for real. They invaded Belgium and Holland to avoid the French defensive position of the Maginot Line and in a lightning Blitzkrieg pushed all before them. In under two weeks Hitler's Armies assisted by Goring's Luftwaffe, achieved what another German Army had failed to do in four years between 1914 and 1918.

For the pilots in 74 Squadron and for all the other air force people in Britain, the beginning of the Blitzkrieg heralded at last the call to action. The phoney war was finally over. Yet for Malan and the others there was still some waiting to be done. Air Chief Marshal Dowding was not willing to fritter away his all too small fighter force in France, especially not his precious Spitfires. Soon enough, however, the British forces in Northern France and in Belgium were pushed back to the Channel coast around Calais, St Malo and Dunkirk.

The waiting and the frustration was over at last. It had started. The legend of Sailor Malan was about to begin.

CHAPTER THREE

Action over Dunkirk

The squadron began to fly operational patrols over the Channel off the French coast as the situation worsened. Malan was usually flying Spitfire K9953 at this time, although a patrol over the Belgium coastal town and port of Ostend on 19 May saw him in L1084. 'Steve' Stephen recalls that on the first patrols over Dunkirk they were detailed to fly patrols at between 6,000 and 7,000 feet but found no sign of the enemy. So Malan decided to raise the patrol height to 8,000 feet but by the time the squadron got over there, he had led them up to above 10,000 feet. This height made all the difference for on the 21st Malan finally met the enemy.

It was early evening when Malan led his A Flight on patrol, the Spitfires leaving the ground at 5.37 pm. He led them up above 10/10ths cloud, the top of which was at 15,000 feet, then on upwards to 21,000 feet. At that height the sky was clear. Behind him flew Johnny Freeborn and Pilot Officer Bertie Aubert as his Red Two and Red Three. Yellow Section, Tink Measures, Pilot Officer Don Cobden and Pilot Officer Ben 'Paul' Draper flew nearby.

The Spitfires cruised above the North Foreland, Malan's keen eye making out the dark stains of bursting anti-aircraft fire about ten miles to the south east. He reported a 'Tally-Ho!' to the Controller back at base, then turned towards France. As they headed out across the Channel they began to get out of R/T touch with the Controller who therefore ordered Malan to proceed on course for ten minutes and, if not engaged by then, to return to base.

Malan ordered his men into the much practised line-astern formation and carried on towards Dunkirk. At 7.20 pm he saw two twin-engined aircraft flying about 200 yards apart. This moment was the culmination of all his training, the moment he had been heading for inexorably since he had joined the Royal Air Force, since he left the sea, since he left South Africa – since his life began. His training took over. It had been practised long enough and the situation was clear. Malan made the obligatory Number One fighter attack as laid down. He selected the right hand machine which he now identified as a Heinkel 111 K, Tink Measures and his section going after the second bomber.

Yet despite all his training, Malan came in much too fast and his first attack overshot. Side-slipping, the Heinkel went ahead again and Malan began firing at a range of 150 yards, closing to 50. His incendiary ammunition began to slash into the starboard wing, the South African clearly seeing his fire entering behind the engine, wing-root and then the fuselage. The under-carriage legs dropped down suddenly. Closing in again Malan began firing, seeing his incendiaries chewing up the German's fuselage. The Heinkel was only 200 feet above a cloud layer but the pilot made not the slightest attempt to drop down into its protective opaqueness. There was no return fire either from the gunner's position atop of the fuselage, Malan assuming his bullets had already put him out of action.

As Malan pulled up and away from the crippled bomber he saw Measures firing at the other Heinkel. Then he spotted more aircraft about five miles away to the east of his own position. Calling his section to follow him, Malan dropped down into the cloud layer, heading towards the other aeroplanes, climbing just out of the cloud for brief glimpses of the enemy machines. As he did so he repeatedly called to Measures over the R/T, giving him the course he was flying (120° magnetic) but the R/T reception was very poor and Measures failed to receive the messsages.

Suddenly Malan and his two wingmen flew into a clearing in the clouds, a fluffy white valley some two miles across opening out beneath them. In the clear space Malan came closer to the aircraft he had been painstakingly stalking. About a mile ahead were five Junkers Ju88s with a sixth Ju88 about 1,000 yards to the left in clear air. Using hand signals due to his defective radio, Malan sent Aubert after this latter Junkers while he and Freeborn again dropped down into the cloud, emerging moments later beneath one of the main bunch of bombers. He opened fire from below at 200 yards range, then throttled back to open fire again, this time from dead astern at 150 yards. The rear gunner in the 88 returned fire, but Johnny Freeborn saw this fire cease after his leader's first long burst. Malan's incendiaries sparkled all over the German's fuselage and tailplane. As he fired again, the port engine blew up and the tail area burst into flames.

Leaving the cripple, Malan radioed for his section to reform and turned onto a westerly course. Only two minutes later Malan saw another Ju88 below. Malan half-rolled onto its tail, commencing fire at 800 yards from the German's starboard quarter. Closing right in he fired again, this time from 400 yards, then his guns fell silent – out of ammunition. He saw his fire hit the 88's starboard wing-root, but then the 88 went into cloud. Meantime, his wingmen had seen two more Ju88s and had attacked. Malan called them up to reform, but Bertie Aubert failed to respond. He was last seen by Freeborn still attacking one Ju88 which was leaving a trail of smoke.

Back at base, Malan discovered that Measures had probably destroyed the second Heinkel and Draper confirmed seeing Malan's Heinkel going down with its wheels dangling down, trailing black smoke while wobbling about violently. Malan was credited with one Ju88 destroyed, the Heinkel probably destroyed and one Ju88 probably damaged. Johnny Freeborn had also destroyed a Ju88, but Aubert was missing.

It later transpired that Aubert had continued after his Junkers and shot it down but was forced down at Berck sur Mer when his fuel ran low. He went by road to Boulogne to try and get some petrol but on the return journey ran into German tanks, so had to turn back, abandoning all hope of getting back to his Spitfire. He made his way to Calais, hitching a ride back to England on the 23rd in a Blenheim.

Everything had come right for Acting Flight Lieutenant A.G. Malan. The years of training had paid off in his very first fighter engagement. Nobody doubted that with his expert gunnery scores during Annual Camps and air firing in general that if he ever got amongst enemy aircraft he would hit his man. If his equally good flying skill and his eyesight helped, in addition, to keep him alive in his first air battles he must soon prove a deadly air opponent for the German Luftwaffe.

> I'd tasted blood at last (he recalled later). The release from tension was terrific, the thrill enormous. I'd been wondering for so long – too long, how I'd react in my first show. Now I knew. Everything I had learnt had come right. There was hardly time to feel even scared.

This was the cold analysis of his first combat. For the first time he had fired his eight Browning machine-guns in anger. He had survived the battle and possible death. Like fighter pilots before him and those who would come afterwards, he felt that first flush of success. Yet he was human enough to confess he had known fear. As Alan Deere told this writer, anyone who said he felt no fear in action was not telling the truth. Youth and inexperience could possibly dull the senses but one quickly discovered that in war it was not all one-sided. The Germans were doing their level best to try and kill or maim you.

Seventy-four Squadron were at dawn readiness the next morning, 22nd May. Malan led three sections off soon after 5 am, climbing away from Rochford towards Dover, levelling out at a patrol height of 12,000 feet. At 5.45 am Malan's keen, clear eyes picked out another aeroplane, a German Ju88, steering a north-easterly course. The Junkers was in an area of clear air between clouds. Malan quickly ordered his pilots into line astern formation and led them on a course to cut off the German's escape route to the clouds. The German crew were alert, however, saw the approaching Spitfires, and the enemy pilot put his aircraft into a dive towards the sea below. The bomber went down steeply; the following Spitfire pilots registered 400 mph IAS (Indicated Air Speed) on their instruments. They also saw four bombs being jettisoned by the Junkers.

Malan closed to 250 yards, carrying out a Number One attack, firing two two-second bursts; the second stopped the return fire from the rear twin-guns. The German pilot knew he was in trouble. With his rear gun silent and bullets hitting his machine he

74 Squadron at Rochford just prior to Dunkirk. Seated l. to r. Bertie Aubert, S/Ldr F L White (CO), A H Smith, Tink Measures and Sammy Hoare. Standing l. to r. Don Cobden, Paddy Treacy, Ernie Mayne, F/Sgt Llewelyn, John Mungo-Park, Johnny Freeborn and Tony Mould.

Malan and his Tigers, standing in front of Sailor's Spitfire "A". l. to r. (?), (?), Paddy Stevenson, Don Cobden, Smith, Malan, Mungo Park, Johnny Freeborn, D Hastings, Sergeant Kirk, Ernie Mayne and Bill Skinner Manston, July 1940.

must act quickly. He tried violent evasive action, skidding and turning from side to side. Yet Malan's deadly fire was hitting the diving twisting target; he could see it blasting the port engine and punching holes all over the fuselage. Already twin ribbons of white vapour were streaming from both engines. The vapour, probably glycol engine coolant, covered Malan's windscreen, forcing him to break away after his fifth burst. The Ju88, he could see, had started to level out and its speed had been greatly reduced. Malan's Number Three, Sergeant Tony Mould, whose radio had failed, followed up his leader's attack firing from 200 yards, emptying his guns. Pilot Officer Peter Stevenson, Red Two, then attacked but following his first burst the 88 suddenly lost height, both engines stopped, and it smashed into the sea and broke up. Two seconds later there was nothing left on the surface except for one empty dinghy. The Spitfires circled, looking for any sign of survivors but nothing was seen. The bomber had gone in ten miles north of Calais.

During the attack, Tony Mould, following Malan, had seen the Junkers begin to level out as the Spitfires came round into line astern of it, and clearly saw too the top gunner firing at Malan's machine. Mould discovered on landing that one bullet had gone through his own port wing tip. As he finished his ammunition, Mould estimated the 88 was down to 280 mph and right 'on the deck'.

It was a satisfactory team effort and the only victory of the day for the Tigers. Their next victory came just over twenty-four hours later, Squadron Leader F.L. White and Flying Officer Measures shooting down a Henschel 126 light observation aeroplane. However, the Henschel's gunner managed to put a bullet into White's radiator, forcing him to make a landing on the airfield at Calais Marck. With the Germans advancing rapidly on Calais it was very obvious that the CO must be rescued quickly. Malan arranged a rescue mission, having been on the same patrol and seen White going down. However, the Station Commander at Hornchurch decided that 54 Squadron should affect the rescue by flying over a two-seater Magister to Calais, escorted by two Spitfires. 54's Flight Lieutenant James 'Prof' Leathart flew the two-seater, and Pilot Officers Alan Deere and Johnny Allen escorted him in their Spitfires.

Malan was a more than a little put out, feeling strongly that it should be the Tigers who should rescue their own CO. In the event, 54 did an excellent job, Leathart bringing White back while Deere and Allen outfought several Messerschmitt 109 fighters, shooting down three for certain, and got themselves home safely. A Blenheim flew two of 74's ground personnel over to Calais in order to repair the downed Spitfire. They were Corporal Higginbotham and L.A.C. Cressay. Unhappily they were both taken prisoner by the too rapidly advancing Germans.

If 74 was not allowed to fly the rescue at least they would go over as top cover. Steve Stephen recalls Malan briefing the pilots saying: 'If anyone suddenly decides his engine is running rough on this show, or turns back for any reason, he'll have me to deal with!' And they knew he meant it. Today James Leathart remembers:

> To me, Sailor was a distant character – I didn't know him very well, I rather doubt if anyone else did. He was older than the rest of us, was married, and lived out for one or two of the peacetime years, so was not in the mess very often.
>
> I only remember him as a very good shot at our peacetime practice camps, much better than the rest of us. He was not, as I remember it, a very good pilot, but if anybody got in his sights, they'd had it.

Paddy Byrne was also lost on the 23rd, being shot down by ground fire. Malan was leading the patrol, Byrne leading the second section. As the Spitfires patrolled near Clarmarais Wood, Byrne's machine was hit, going down with smoke pouring from his shattered engine, and he himself being wounded in the leg. He too was taken into captivity, ending up in Stalag Luft III for the rest of the war. He died in 1979.

In many ways it had been a frustrating day.

The 24th was to prove little better, although at least the Tigers managed to hit back and gain several victories. Bertie Aubert returned after his forced landing on the 21st, and was raring to have another crack at the enemy. He flew on one of the squadron's patrols twenty miles south of Dunkirk, but this time he did not get back. Others lost on the 24th were Sergeant Tony Mould, Flying Officer Sammy Hoare (in K9321), and Flight Lieutenant Paddy Treacy went down. In addition to these three missing stalwarts, Mungo-Park returned to Rochford with slight flesh wounds to his left arm. Mould got back by boat and so did Treacy, but much later. Hoare force-landed at Calais Marck where he found the two missing ground crewmen but like them he was later captured.

Sailor Malan was engaged in the late morning. With the evacuation of the BEF in full swing, the squadron was constantly flying patrols over, in front of or to the rear of the beaches. Below, lines of soldiers could be seen standing at the water's edge taking their turn to wade out to be picked up by the myriad of small craft that plied to and fro from larger vessels off-shore. Only the often sudden appearances of German aeroplanes would send the troops scattering into the sand dunes or to any other cover that might offer even the slightest shelter from bombs, cannon or machine guns.

Number 74 Squadron was on patrol off Calais. Smoke was rising into the air and far below the British pilots could see the mass of tiny ships off the coast or far out into the Channel. Then, five miles or so out to sea and below a cloud mass, Malan spotted a Dornier 17 bomber. It was obviously looking for shipping to bomb, but made off rapidly at high speed when its crew saw the Spitfires.

Malan was after it in an instant but the Dornier's pilot took advantage of the 8/10ths cloud and easily slipped into it. Yet there were gaps in the cloud and whenever the Dornier reappeared, Malan got off quick bursts at it. The majority of these bursts were fired at from between 400 to 500 yards because he was finding it difficult to close in quickly each time the Dornier came into the open. Finally his eight guns were exhausted of ammunition but not before the bomber's starboard engine was set on fire and pieces had been shot away from it. It dropped below the cloud, and the rest of his section then fired at it. Then the Dornier burst totally into flames and crashed.

Malan had followed below the cloud and filmed the crash with his gun camera. As he watched, he saw what he took to be the pilot scramble from the burning wreck and pull out another crew member.

Later that afternoon the squadron was again off Dunkirk. Malan was leading Yellow Section of just four Spitfires on an offensive patrol over Dunkirk-Calais-Boulogne, when he spotted bursting AA fire above, at 12,000 feet. The Spitfires were just off the coast west of Dunkirk. Malan put his men in line astern and climbed up, keeping his eyes peeled as he began to investigate. Then he saw them: three vic formations of Heinkel 111s above, the first vic consisting of nine aircraft, the second of twelve and the third of nine. Malan decided to attack the middle group of twelve, although the Spitfires had to fly through very heavy and accurate anti-aircraft fire – Allied anti-aircraft fire!

Malan picked the starboard flank, taking his men up in echelon port formation from astern of the Heinkels. However, the bombers had an escort of fighters, Messerschmitt 109s and 110s, which could be seen above in the sun, turning in preparation for an attack. Malan could count at least eight Messerschmitts but suspected there was more above in the haze. He estimated there was still time to make a pass against the bombers before the Messerschmitts could come down on them. As he manoeuvered into his attacking position, one Heinkel was hit by the AA fire and began dropping away from the formation with its port engine stopped and one wing well down. Malan continued to climb and attacked his selected Heinkel, giving it three quick two-second bursts at both engines and fuselage, firing from 250 to 150 yards.

Then there was a bang. His Spitfire shook and a hole appeared through its starboard wing, and he felt a jolt somewhere in the fuselage. He had been hit by 'friendly' AA fire. It severed the electrical leads near his seat and the light in the reflector gun-sight went out. He pulled up and looking back he saw a Messerschmitt 110 closing in on his right and a 109 right behind. This was no time to be hanging about. He pulled his Spitfire into a number of very steep climbing turns into the sun which successfully threw off his two unwanted antagonists. What did please him, however, was the sight as he was making his steep turns, of his Heinkel going down, and its crew taking to their parachutes.

Able then to take stock of the situation he attempted to get his gun-sight back into operation. He changed the light bulb but to no avail; Malan concluding that it was the wiring that had been damaged. Looking about he found the battle had moved on, or just that he had flown out of it. As his fuel was running low it was obvious that he had little time to try and locate the action so he flew home.

Due to heavy cloud formations early on the 27th, 74 Squadron flew at varying heights between 2,000 and 15,000 feet on a patrol line Calais to Dunkirk. Just on 9 am bursting anti-aircraft shells attracted Malan, flying as Red One, again in K9953. The dark stains blossomed at around 5,000 feet to the south of Dunkirk. Leading his pilots to investigate, they flew around huge mounds of white cumulus clouds but failed to find any trace of hostile aircraft.

Malan turned north towards Dunkirk itself. It was hot in the cockpit. The May weather was proving glorious but terribly hot, especially under the perspex cockpit hood. It foretold of a hot, sunny, blue-skied summer.

Nearing Dunkirk several Messerschmitts could be seen flying above and behind the Spitfires. Malan ordered the squadron into pairs, and he and his Number Three pulled up towards a 109 directly above. Still in a nose-up climb. Malan gave the single-seater two full deflection bursts from 100 yards range. The 109 quickly dived towards the clouds, Malan following, firing four more bursts from around 300 yards, but being unable to close the gap between himself and the diving Messerschmitt. But before the German reached the clouds, heavy smoke began to pour out of the starboard side of the engine cowling. Malan's fire had taken deadly effect.

The action ended at 9.15. Malan turned for home, meeting up with Red Four, Pilot Officer P.C.F. Stevenson. Both pilots were now low on petrol when they saw a group of eight Dorniers flying in two vic formations of five and three. Despite their fuel state they attacked the rear vic of three, Malan taking the leader from the right flank, Stevenson attacking it from the left. The Dornier's rear gunner began to return fire as the two Spitfires closed in,

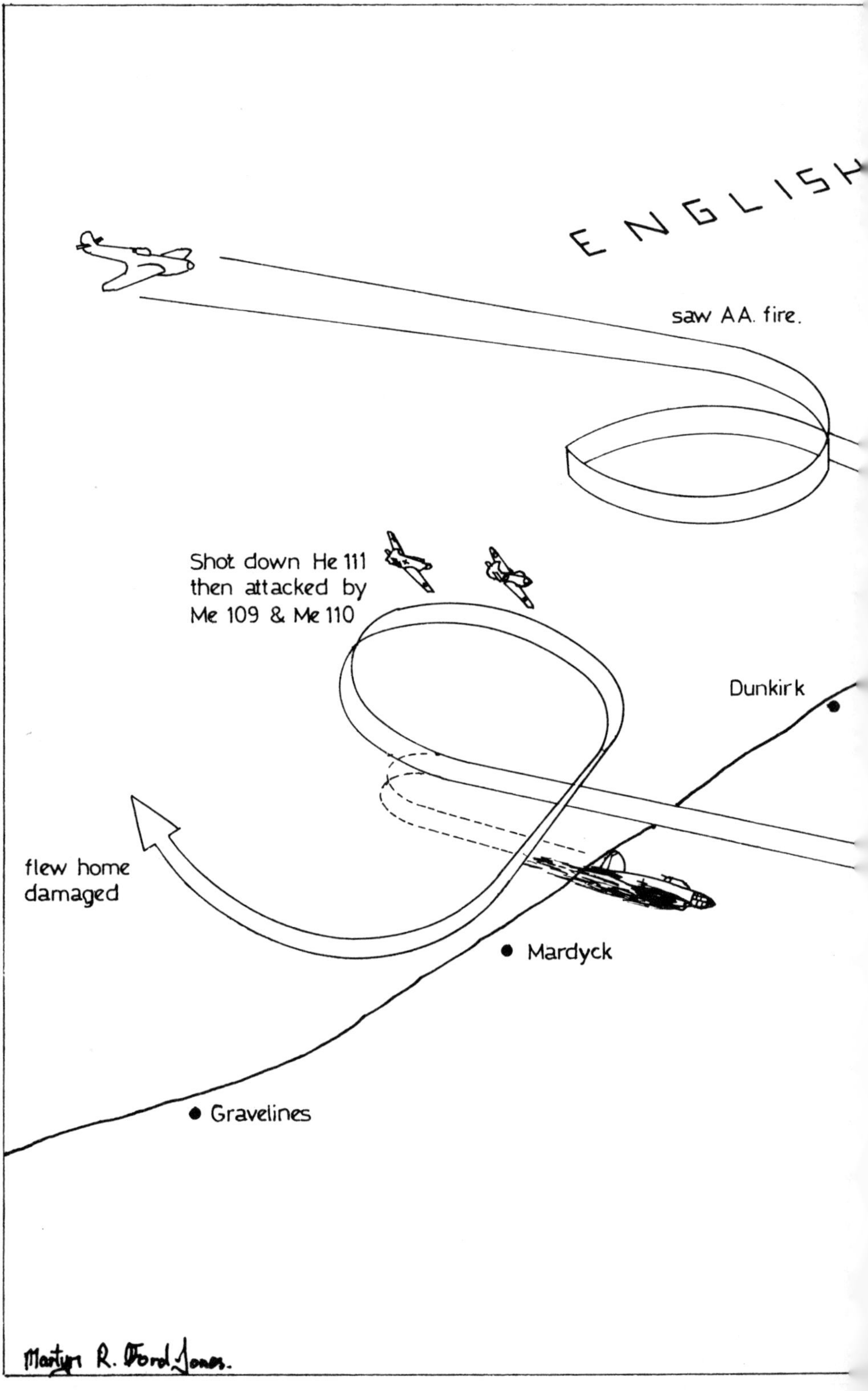
ENGLISH
saw AA. fire.
Shot down He 111
then attacked by
Me 109 & Me 110
Dunkirk
flew home
damaged
Mardyck
Gravelines
Martyn R. Ford-Jones.

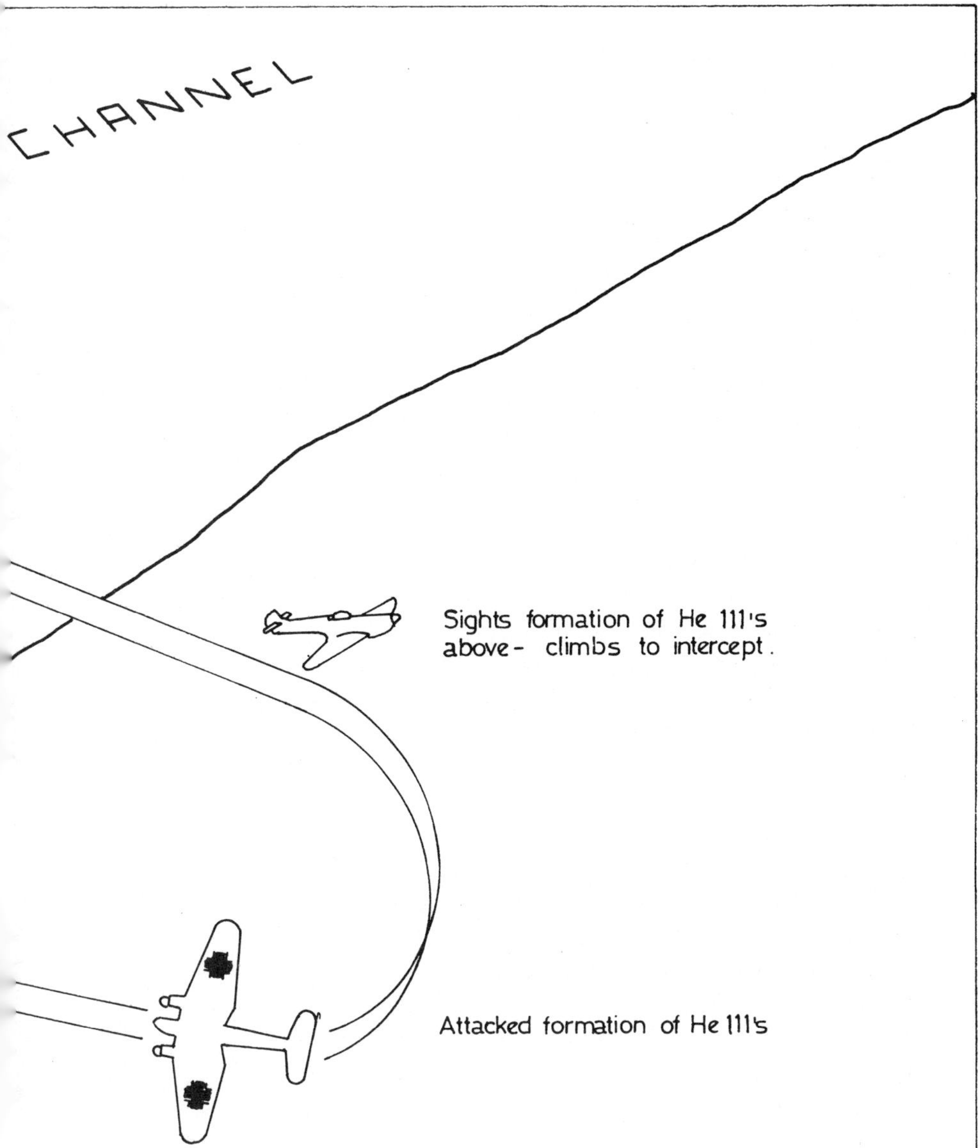

24th May 1940
Patrol.
Dunkirk - Calais - Boulogne

while his pilot headed for the clouds and a German AA barrage. As they could not afford a long chase with rapidly drying fuel tanks, Malan and Paddy Stevenson closed quickly. Malan fired four quick bursts from 300 to 100 yards, aiming for the Dornier's port engine and fuselage. Breaking away to the right, he glanced back to see smoke coming from the Dornier's tail and port engine. Coming round for a second pass, he saw Stevenson in difficulty, having apparently been hit in his glycol feeder tank. He would need to get down quickly before his engine overheated, siezed up and stopped. Leaving the Dorniers, Malan followed Stevenson, calling him over the R/T but receiving no reply. Stevenson suddenly changed course from cross-Channel to dive towards Dunkirk. Malan assumed he preferred an attempt at landing on the beach rather than being left with a dead engine in mid-Channel. Malan could not afford to escort him, and reluctantly had to pull round and head back to England. He landed back at Rochford with just two gallons of petrol left.

It had been a reasonably successful morning. Five Me109s had been claimed as destroyed, plus two more probably destroyed, and Malan had damaged one Dornier. Only Stevenson (L1084) was missing and he was probably down somewhere around Dunkirk.

There was more action in the late afternoon, when 11 Spitfires took off at 3.15 pm. They arrived over the French coast but saw nothing. Half an hour later, at around 4 o'clock, Malan led the patrol down to a lower altitude. There they saw several groups of German bombers, again in formation of five and three heading for Dunkirk. The squadron split up to deal with them. Malan and his Red Section divided into pairs, and went down towards a wide vic formation of five. These bombers were Dornier 17s, and they began to split up also, each diving away towards the south-east. Malan and Red Three, Pilot Officer Hugh Dowding, carried out a Number One attack upon one Dornier, sweeping down at high speed and firing from 250 down to 50 yards. The three aircraft were now down to tree-top height, and Dowding made the last attack at only 25 feet above the ground. By this time the Dornier's port engine was burning and smoke was coming from the rear of the fuselage. The two Spitfires, now quite some distance inland, near St Omer, came under severe and accurate light ground fire, forcing them to break off the chase and return to the coast, being almost continually under fire all the way.

Back at Rochford at 4.45 pm, Malan waited for B Flight to return. Paddy Treacy (in K9875) had led his men over five minutes later than Malan and had also become involved with Dorniers. As most of 74 began to drift back, they added up the score. Tink Measures and Peter St John had got one, Skinner another. Steve Stephen landed. He reported that he and Treacy had shot down a third Dornier but that then his leader had disappeared. Nobody worried over much. Treacy had got back before, he'd soon find his way home again.

At the end of this busy day, 74 Squadron was pulled out of the line and sent north to Leconfield. Over the Dunkirk evacuation, which was still far from over, the squadron had lost one pilot killed, three taken prisoner, (Paddy Treacy was not so lucky this second time and was captured) and one slightly wounded. Other missing pilots had showed up. Paddy Stevenson too got back and joined the Tigers at Leconfield.

It was time now to rest briefly, take on replacement pilots and also to re-equip with some promised Spitfire Mark IIs. Malan, ever the burning light, could see no need to take a breather and train new pilots, although he himself could see the merit of a rest he was eager to continue the fight. It had been a long time coming and now they had been pulled back after only about a week in action. He need not have worried. 74 would soon see action enough. The leader of A Flight had proved his ability in combat, for in those few days above the greatest military evacuation in history, he had destroyed three German aeroplanes, helped to destroy two more and damaged or probably destroyed four others. It was no longer ground targets or firing at towed drogues; this was the real thing. His shooting was just as good if not better.

The promised re-equipment of Spit IIs did not happen before the squadron returned to Rochford on 6 June. Added to that disappointment was the fact that the evacuation from Dunkirk was over. There was still fighting going on in France but it was all too clear that France must eventually fall and leave England alone. Then the Germans must turn their full attention to Britain.

Meanwhile it was a period of preparation for what must come, preparations partly based on evaluation, evaluation of what they had so recently gone through and had learnt over France. An early change which Malan had made was the formation in which the squadron flew. Throughout his RAF service, the peacetime air force had flown set formations, laid down and honoured by standing orders. In Fighter Command the basic element was a section of three, four such sections totalling the normal twelve-man squadron in the air. The main flaw in the three-man section was that the Numbers Two and Three were kept so busy flying formation with their leader that they had no chance of keeping a really good look-out for enemy aircraft. And if the truth were known a few section leaders might be watching too closely, perhaps, an inexperienced wingman in case he should edge in too close and chew off his own tail! Anything that kept the eyes from all parts of the sky was detrimental to staying alive in the air.

The German pilots had already developed a two-man section element in Spain and had been and were using it to advantage in Poland and France. Malan quickly realised that three was wrong, two was right. Not that he had necessarily seen the Germans were flying in pairs or fours over Dunkirk. In that battle the fights had been too disjointed. Rarely did one have the opportunity of looking at what German fighters were doing; it was really a case of getting to them and mixing it if one could.

The other fact that emerged from their first encounters was something that the air fighters had found in the Great War, and that was the advantage of height. Malan had found the enemy patrols above 12,000 feet, and by the later stage of 74's Dunkirk actions they were invariably themselves patrolling at heights of 20,000 to 24,000 feet. This was one answer as to why the Army was saying, 'Where was the RAF at Dunkirk?' They were there, but often so high that they could not be seen, nor heard above the sound of gunfire on the ground. On other occasions the RAF were fighting inland or out to sea above the ships.

Numbers 54 and 65 Squadrons, still at Hornchurch, were also aware of Malan's tactics. Indeed both Bob Tuck and Al Deere recall discussing these ideas with Malan and each other and their other flight and squadron commanders. Tuck was now with 92

Squadron as a flight commander but he too could see the disadvantages of the three-man section and had also changed over to a four-man section during the Dunkirk operation, being able then to split it down to pairs very easily. Tuck recalls:

> Ninety-two had lost six pilots in two days, including the CO and the other flight commander. I figured it out darn quick that the 109s were coming at us 'woomph' – just like that, all in loose formations and we were flying all this jammed-in, odd numbers, which was absolutely hopeless. No freedom of action at all. You were concentrating because of the possibility of collision when you should be looking around at what was going on everywhere else. I eased them out into pairs when I took command. Copy-book formation stuff was hopeless.

So, as can be seen, the pilots of the calibre of Malan, Tuck and Deere, were all flying their own styles and beginning to think for themselves, to change the accepted rules now that the game was starting to be learnt. They had also changed the accepted thinking on gun harmonization. With eight guns in the wings of the Spitfire or the Hurricane, the official pre-war thinking was to fill the air with .303 bullets around 400 yards ahead of the line of flight with a sort of shot-gun blast effect. Each gun would be individually sighted against a large pattern set out on the side of a hangar. The idea was that a percentage of shots would be bound to hit the target aircraft.

Malan and Deere at Hornchurch had already discussed harmonization. Malan, whose pre-war shooting ability had been confirmed by his recent successes over Dunkirk, was adamant that his guns and those of 74, would be harmonized so that the bullets would converge to one point, at 250 yards ahead of his Spitfire. It was this 'area of lethal density' which Malan and others like him were to prove so deadly.

Malan, Deere, Tuck and others were not alone in insisting that harmonization at 250 yards was the most effective, and eventually Fighter Command bowed to the pressure of the men up the sharp end, and issued instructions that this should be standard for all day fighters. The great secret of success, however, was to get in close – so close that one couldn't miss!

Throughout the Dunkirk operation most fighters, including those from Hornchurch, had not been armoured. 74 had suffered several casualties during the operation and it was felt that some might have been avoided had armour been installed. Yet better late than never, and as the squadron returned to the south, all had armour plate behind the pilot's seat and a bullet-proof windscreen in front of them. The windscreen and engine gave good protection when attacking bombers. Tuck's pithy comment sums it up but also shows one weakness – to cross-fire:

> Cross-fire when attacking bombers. Dirty great engine in front and thick windscreen but they were spread out and we were vulnerable from the side. In close, watch out for debris coming back, it just whipped back and of course his oil on your windscreen and you're buggered!

With Dunkirk over and France in its last days of active resistance, the Royal Air Force and especially its fighter pilots, stood ready to oppose the onslaught against Britain. The Few were about to have their day.

CHAPTER FOUR

The Vicious Sky

On 1 June 1940, before 74 Squadron returned to Rochford, Acting Flight Lieutenant A.G. Malan was awarded the Distinguished Flying Cross. The citation to this his first decoration, and a very popular one with the Tigers, recorded that he had led his flight, and on occasions the squadron, on ten patrols over the Dunkirk evacuation. It went on to record his ' . . . great skill, courage and relentless determination in his attacks upon the enemy.'

Back at Rochford once more, 74 Squadron awaited developments. The sky over the English Channel and southern England was still largely clear of any hostile actions by the Luftwaffe, save for the occasional incursion of lone reconnaissance flights by Dorniers or Junkers 88s. Yet there was growing activity at night during this period of calm before the storm that must surely be on the way.

Malan, ever aggressive and ever anxious to get to grips with the enemy, had 74 Squadron flying increasingly at night. Whether it was practice flights or operational sorties, all had to take an active share. Nobody liked flying at night and the Spitfire was not the ideal machine to fly at night, especially with its long engine cowling right in front of the pilot and its narrow space between under-carriage wheels, but they did it uncomplainingly.

Then on the 13th Sailor went down to the Royal Aircraft Establishment at Farnborough. He was one of several operational pilots asked to fly and help evaluate a captured Messerschmitt 109E-3. This particular Messerschmitt had been with Jagdgeschwader 54 (JG54), but its pilot had lost his way and believing he was on the German side of the border had landed intact in an orchard at Woerth, Bas-Rhin, on 22 November 1939. Following a period at the French Test Centre at Orleans-Bricy, it was handed over to the British at Amiens on 2 May 1940 – just eight days before the German offensive began.

On 2 May, the 109 was flown in mock combat with Flying Officer M.H. Brown, a pilot with 1 Squadron in France, and the next day, this same pilot, Mark 'Hilly' Brown, flew the 109 to the A & AEE at Boscombe Down in England. Then it went to Farnborough on 15 May. Here it was test-flown by RAE test pilot Flying Officer J.E. Peabody and also by operational pilots. On 8 June Bob Stanford Tuck and Flight Lieutenant G.H. Stainforth flew the Messerschmitt in action against each other, taking it in turns to fly a Spitfire while the other man flew the 109. On the 13th Malan flew the German machine.

There can be little doubt that Malan learnt as much as Bob Tuck from being able to sit in and fly the aeroplane that was their most dangerous adversary. They had already met it in actual combat over Dunkirk and both would meet it on many occasions in the months to come. Tuck recalls: 'It was a tremendous advantage having sat in the thing and flown it. You somehow felt you knew it and even when you were attacking one in the air, you had a mental picture of what he'd do.'

Two days after this 109 flight, Tink Measures was confirmed as the unit's B Flight Commander, having taken over the post following the loss of Treacy. It was the same day that news arrived that Paddy Byrne was alive and a prisoner of war.

The squadron was still flying at night and standing-by for action should German night raiders be reported. Malan, always the task-master, naturally took his share at readiness during these duties.

On the night of Tuesday 18 June, Malan had vastly different thoughts than nocturnal raiders, for he had recently become a father. His wife Lynda had presented him with a son, whom they named Jonathan, born in a nursing home at Westcliffe-on-Sea. (1) However, the war suddenly intruded upon his thoughts and upon his general well-being. Air raid sirens started to wail as the day began to end. Malan, still fully dressed, went outside. It was a pleasant mid-summer night with clear skies and a nearly full moon. Towards the coast in the direction of Southend, searchlights were

(1) Although Sailor was always called 'Sailor' by his friends and the public in general, he was always known as John by his wife and more intimate friends.

probing hopefully across the heavens. Bombs were falling and Westcliffe was near to Southend.

Malan received permission from Squadron Leader White to get airborne, his rigger and fitter turning out in pyjamas, gum-boots and steel helmets to start his Spitfire. The three men drove to the dispersal area where the two airmen started up the fighter while Malan strapped on his parachute. Malan climbed into the cockpit of K9953, the engine already purring sweetly. As the engine warmed and Malan prepared for take-off, he could see one German aeroplane caught in the searchlights at around 6,000 feet. Even as he watched he could see it was heading directly for the airfield. Leaving the cockpit a good deal faster than he had climbed in, he dived into a freshly dug slit-trench. Earlier in the day this trench had only been 18 inches deep but unknown to him someone had completed it and he fell headlong into a five foot void, landing heavily in the muddy bottom just as the German roared overhead.

Climbing out of the trench he quickly jumped once more into the Spitfire and seconds later was climbing hard after the German machine which was fast disappearing. It was now twenty minutes into the morning of 19 June.

The enemy pilot was heading for the coast, still held in the beam of a searchlight, flying at about 8,000 feet. Malan caught up with the bomber, which he could now plainly see was a Heinkel 111; he positioned himself directly behind it, commencing to fire at 200 yards, and letting go a four-second burst as he went in to 50 yards. He could clearly see his bullets hitting the Heinkel, but then his windscreen was covered with oil, forcing him to break away to the left and below the bomber, seeing it from this angle begin to spiral out of the searchlight's beam.

As his windscreen cleared sufficiently for him to see ahead, he saw another German aeroplane held in searchlights to the north of his position. Closing with this one, Malan fired from 250 down to 100 yards, giving it five two-second bursts, taking care not to overshoot the target. Bullets splattered and sparkled all over the luckless bomber as it turned to the left. Already it was trailing thick smoke and then a parachute suddenly opened close by. The machine, it was another Heinkel, went down in a spiral dive still held in the searchlight and followed by Malan, until it crashed in flames very near to Chelmsford. As he circled the crash he switched on his navigation lights for a short time to help establish his identity to the anti-aircraft defences and as he turned and flew home, gave the letter of the day when at 3,000 feet as the searchlights appeared to be trying to locate him. At the beginning of both attacks, Malan had taken time to flash a succession of downward recognition lights before closing in, in order to tell the ground defences that he was attacking. Surely a sign of the professional, unhurried and unruffled in the heat of action.

He landed at 1.30 am. Tink Measures had also gone up in P9398 at 00.50 am, landing also at 1.30, but he returned empty handed. Malan's first victim had crashed on the coastal beach, half in and half out of the water. The second Heinkel, held by the lights all the way down, shedding bits and pieces as it fell, crashed at Chelmsford. Eye-witnesses, many having seen the whole action and heard the rattle of machine-guns in the night sky, saw the bomber narrowly miss a house before smashing through a fence and some trees, then burst into flames. There was a momentary pause followed by a tremendous explosion as the Heinkel's petrol tanks exploded, the sound being heard as far away as Southend.

It had actually crashed into the garden of the Bishop of Chelmsford, Doctor Henry Wilson, and the wreckage blocked Springfield Road. The people watching on the ground cheered loudly as it came down. A pall of smoke rose above the crash site and the watchers distinctly heard the sound of Malan's engine as, in their words, his Spitfire '. . . dipped in salute over his fallen enemy.'

Two of the crew were found dead in the charred wreckage; the rear gunner still strapped at his post, was also dead. These three were Oberleutnant Heinz Corpus, Oberfeldwebel Grosz, and Feldwebel Vick. The fourth crewman, as seen by Malan during his attack, had baled out, landing near a farmhouse at Writtle. He had been wounded in the arm, so he knocked on the door of the nearby house, giving the lady occupant quite a shock when she found a wounded German airman standing on her doorstep. Regaining her composure she took him in before calling for the police. From her house he was taken to the Chelmsford and Essex Hospital where his wound was treated while under guard. One patient in the hospital related to the local newspaper that the German had been in good humour and had commented that if he had known he was coming to stay he would have brought some soap and a flannel with him. His name was Leutnant S. Erich, aged 27. His arm wound was fixed and he was discharged the same day.

The three dead airmen were buried in Chelmsford Cemetery in Writtle Road, Chelmsford on 21 June; Doctor Wilson conducted the funeral service. The men were afforded full military honours by officers and men of the RAF, their coffins being draped with a German flag as three volleys were fired over them. An RAF tender conveyed the bodies to the cemetery and two wreaths were laid. One was from the RAF, a second from the AFS inscribed 'When duty calls one must obey.'

Two kills at night in a single-engined Spitfire was quite a feat, and it went down well with 74 Squadron. Not too many fathers have such a unique way of defending a wife and son as Malan had done that June night.

One week later 74 Squadron left Rochford to return to Hornchurch, 54 Squadron took its turn at Rochford, and 65 remained at the parent station.

Two days later, 28 June, King George VI paid a visit to RAF Hornchurch, the officers and men turning out to form up near to the parked Spitfires of 65 and 74 Squadrons. Three pilots of 54 Squadron flew up to the parade, Prof Leathart, Al Deere and Johnny Allen, who, with Bob Tuck and Sailor Malan, were to receive decorations from His Majesty. Leathart received the DSO for his rescue flight at Dunkirk, and Deere and Allen collected DFCs for their magnificent escort job. Tuck and Malan were given their DFCs for work over Dunkirk, while the Commander in Chief of Fighter Command, Air Chief Marshal Hugh Dowding, looked on. He knew only too well that it was pilots of the calibre of these five who now stood before their King that would very soon be called upon to even greater feats of daring, while defending Britain in earnest. Perhaps it is the measure of these five that four were to survive the war, only Johnny Allen being destined to die in action less than a month after the ceremony at Hornchurch. And the four others, Leathart, Deere, Tuck and Malan were to destroy and probably destroy more than 100 German aircraft between them by the end of the war.

With the Battle of France over, Britain and her fighter pilots waited. What was later to become known as the Battle of Britain was soon to begin and its opening phase, the fighting over the Channel ports and convoys, was, in the words of H.M. Stephen of 74 Squadron, some of the most vicious fighting he was ever to be involved in.

These Channel battles, for the Tigers, started on 6 July, the same day that the *London Gazette* made Sailor Malan's promotion to Flight Lieutenant (from on acting rank) official.

Flight Lieutenant Measures and Pilot Officer Dowding attacked two He111s which they encountered over Dover. Measures so crippled one that it eventually crashed into the sea near the beach off the French coast. Dowding saw his bullets hitting the other Heinkel but then he lost it in cloud.

Two days later Measures and his section were again in evidence. In company with Dowding and Sergeant Wilfred Skinner, they attacked another Heinkel at 11 a.m. Dowding's fire silenced the rear gunner and following Skinner's attack the Heinkel dived into the sea in flames. In the afternoon four Me109s of 4/JG51 were engaged over RAF Manston. Sergeant Mould forced one to come down at Eltham in Kent, and its pilot, Leutnant Johann Bohn was captured. Paddy Stevenson damaged a second Messerschmitt.

On the 10th, Red and Yellow Sections were scrambled from Manston at 10.38 a.m. The squadron was now using Manston as an advanced base which was to become known as 'Hell's Corner'. At the end of each day the squadron would return to Hornchurch.

Above a convoy code-named 'Bread' off the North Foreland they found a Dornier 17 escorted by at least 30 Me109s. Try as they might the Spitfires could not get to the Dornier, for they were strongly engaged by the 109s. 74 made several claims; Freeborn reported that his Me109 just 'dropped out of the sky', and Stevenson claimed another 109. Four more were claimed as damaged. Freeborn and Mould both landed at Manston with battle damage to their machines.

Eight Spitfires took off at 1.45 p.m. led by Measures, again to provide air cover for the convoy. Also in the air were machines from 32, 56 and 111 Squadrons. A terrific air battle developed, and 74 claimed four Me109s and five Dorniers as damaged. One

Spitfire flown by Pilot Officer D.G. Cobden had to make a forced landing at Manston after taking hits. Malan's chance to score came on the 12th.

On that Friday there was some particularly savage fighting above another Channel convoy which had sailed from the Thames Estuary, code-named 'Booty'. It was a grey, cloudy day, with occasional outbreaks of rain. Throughout the day relays of British fighters protected the convoy and another one named 'Agent', steaming in the opposite direction. 17 and 151 Squadrons had been engaged during the morning but then the weather restricted further action from the Luftwaffe, although from time to time German bombers tried to sneak in an attack. 145, 603, 43 and 501 Squadrons met these later attacks and then Red Section of 74 Squadron was scrambled from Manston at 4.30 p.m. when yet another bandit was reported 15 miles to the north-east of Margate. Out over the sea and amid huge cloud layers, Malan (in P9306) saw exploding anti-aircraft shells from some of the ships and almost immediately he spotted a Heinkel 111 in the process of bombing one of them.

Malan ordered his section into line astern, Sergeant Mould (N3091) and Paddy Stevenson (K9870) swiftly sliding into position as their leader closed in to open fire from 300 yards. The alert rear-gunner started to fire back but ceased abruptly as Malan ended his pass. In turn, Mould and Stevenson attacked and the Heinkel lost height and crashed into the sea. The bomber came from II Gruppe of Kampfgeschwader 53 (KG53).

On 15 July Flight Lieutenant D.P.D.G. Kelly arrived from Number 6 OTU as a supernumerary flight lieutenant. Piers Kelly had got onto fighters by accident in France before the collapse. His usual mount had been the Westland Lysander of Army Co-operation Command but during the German offensive he had discovered, quite by chance, several Hurricanes and Fairey Battles on an air-strip. They had been ferried in by reserve pilots. He informed Headquarters who sent pilots from 1 Squadron over to collect them. Having been quite fortuitously once shown the taps of a Hurricane at Thorney Island on one occasion and then being allowed to fly one round the airfield circuit a couple of times, he managed to wangle himself on attachment to 1 Squadron. Air

Co-op work was by that time at an end in France. He flew with the squadron until they had to evacuate to England. On the squadron's last patrol he was above the converted troopship *Lancastrian* in St Nazaire harbour; the section was led by Sergeant George Berry. German bombers attacked the ship and although Berry shot down one, the *Lancastrian* was hit. She exploded in flames and sank later with very heavy casualties. Kelly later discovered that several of his former unit were on board the ship, ready to evacuate to England.

Returning to England he converted onto Spitfires and then to his great joy found himself posted to 74 Squadron. He was immediately impressed by Sailor Malan who always took great interest in all new pilots. He gave Kelly every bit of advice and showed him the ropes generally which Kelly felt was a great kindness, especially as he was so 'green'.

Kelly found, as Stephen had found, that Malan was a firm disciplinarian. He kept everyone on their toes and never let up on training. In order to get all his aircraft up above cloud and still in formation, Malan would practise flying up through thick cloud in sections, Kelly remembers vividly, climbing hard, his only point of reference being the tail-wheel of the Spitfire immediately ahead of him in the greyness. Yet it worked. Once through the cloud Malan, still leading, was ahead of all 12 Spitfires and ready for action. Both Kelly and Stephen remember Malan practising squadron take-offs in sections – very exhilarating and fun too once they found they could do it.[(1)]

July 19, a Friday, became famous, or infamous, as the day the Defiants of 141 Squadron were massacred by Me109s around mid-day; seven were shot down or severely damaged. Ten of the two-man crew members were lost, and two more wounded. Later in the day Malan led 74 Squadron off to patrol over Deal. German bombers were reported at 6,000 feet with layers of Messerschmitts at between 10,000 and 12,000 feet.

As Malan reached the coastal port of Dover he saw salvos of

[(1)]Kelly's brother Dermit was also an RAF pilot. He flew Battles in France with 104 Squadron. Later, when his squadron converted to Wellingtons he flew more than 20 missions, but failed to return from a raid in mid-1941.

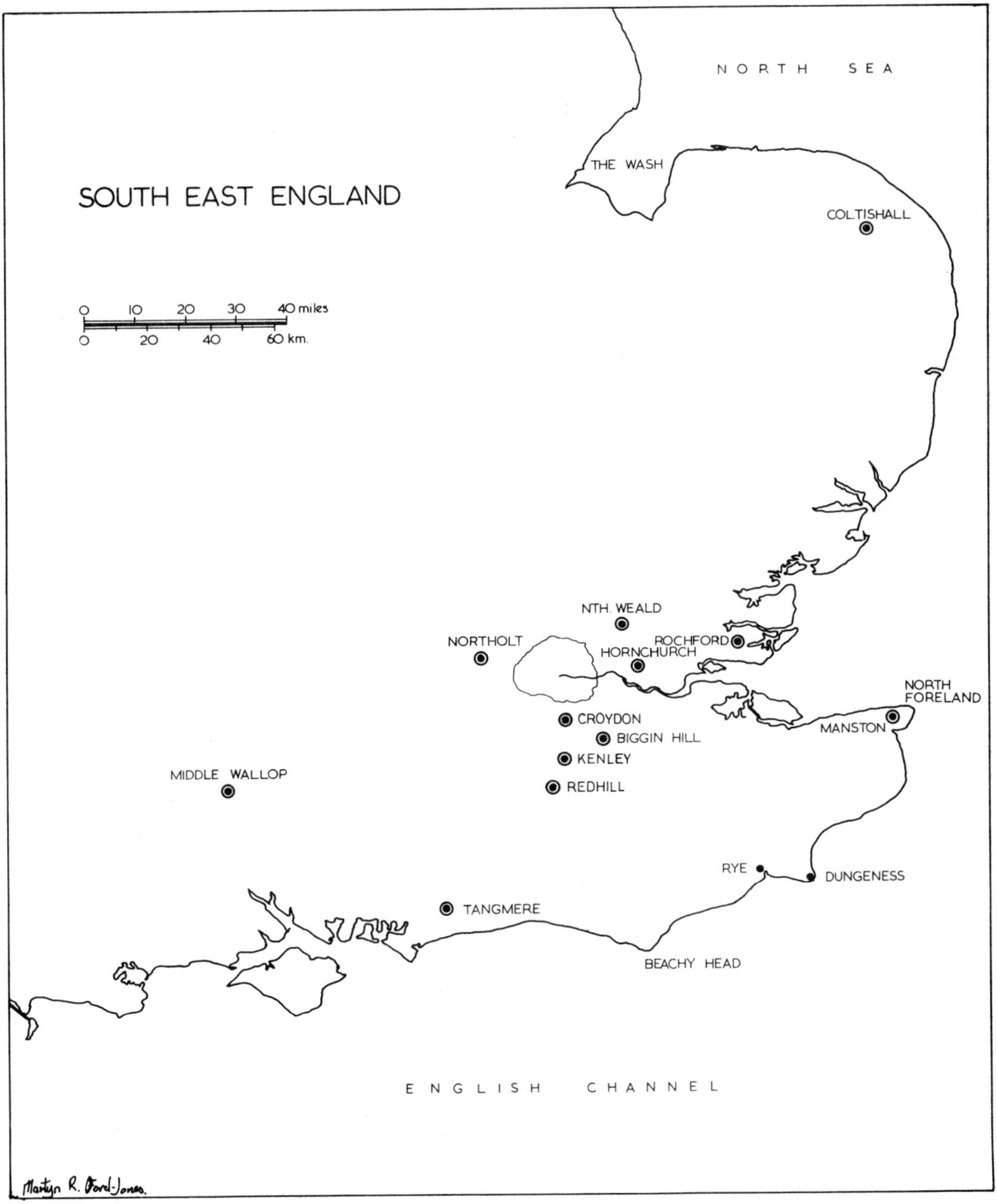

NORTH SEA
THE WASH
SOUTH EAST ENGLAND
COLTISHALL
0 10 20 30 40 miles
0 20 40 60 km.
NTH. WEALD
NORTHOLT
ROCHFORD
HORNCHURCH
NORTH FORELAND
MANSTON
CROYDON
BIGGIN HILL
KENLEY
MIDDLE WALLOP
REDHILL
RYE
DUNGENESS
TANGMERE
BEACHY HEAD
ENGLISH CHANNEL
Martyn R. Ford-Jones.

bombs falling in and around the harbour. At the same moment, Control informed him that the raiders had turned south. Ordering the squadron into line astern in sections Malan began to climb towards numerous groups of fighters that could be seen above quite clearly.

At 13,000 feet, Malan spotted three aircraft and led his section off to the left to intercept. Closing in he saw they were two Me109s in a tight circle with a lone Hurricane. Malan fired a two-second burst at one of these 109s from 100 yards, as it was trying to turn onto the Hurricane's tail. Closing to 75 yards he then fired; he saw the Hurricane break away and Malan applied full starboard bank to turn onto the Messerschmitt's tail. He let go two two-second bursts which caused smoke to stream back from the right side of the 109 as it straightened up momentarily before nosing down into a staggering dive. Looking round he saw the other 109 turning in towards his right, so he climbed steeply to the right, meeting other groups of fighters which proved to be Hurricanes. Looking back and down he saw the 109 being pursued by a Spitfire which turned out to be flown by his Number Three, Paddy Stevenson.

Back at base, Pilot Officer D. Hastings confirmed seeing Malan's 109 diving steeply, wings rocking laterally as if out of control. Malan had fired 1,200 rounds. Stevenson had left the second 109 streaming black and white smoke in a 40 degree dive from 4,000 feet. One Messerschmitt of II/JG51 crashed near Chartres with a wounded pilot, but whether it was Malan's or Stevenson's victim cannot be known. In the event both pilots received credit for a 109 probably destroyed.

The next day two new pilots arrived from 6 OTU, Flying Officer William Henry Nelson DFC, an American who had joined the RCAF and won his DFC in France flying bombers, and Pilot Officer D.N.E. Smith. On the 24th, Flight Lieutenant Measures left the squadron, going to 7 Operational Training Unit as an instructor. Following his departure, Piers Kelly took over the command of B Flight.

On Patrol the next day, Malan led the Tigers and over the coast saw a formation of Hurricanes. Closing with them he then saw several German Me109s heading towards Calais. He managed to get in a quick burst at one from 150 yards, seeing his bullets entering the Messerschmitt's fuselage but he could only claim it as damaged. He was more successful on the 28th, and so was the squadron.

The 28th was a Sunday. 74 flew down to Manston at the beginning of the day to be ready for immediate action. It was a fine day which was a welcome change from the summer storms of the previous few days. With the clearer weather heavy raids were expected; the sector controllers at Biggin Hill, Hornchurch and North Weald moving their squadrons forward to Hawkinge, Manston and Martlesham. Yet they had to wait until after lunch before air activity was finally plotted. It looked as if it was Dover yet again that was going to be in trouble when a large raid began to show signs of approaching the harbour town.

At Manston the telephone rang; its message sent Malan and eleven other pilots racing to their waiting Spitfires. They were airborne at 1.50 p.m., climbing hard in the afternoon sunshine. Below, the sea did not look uninviting but it was the sky above that was being keenly searched by twelve pairs of eyes. There was some cloud at 6,000 feet but the controller informed Malan that the raid was coming in at 18,000 feet. The Spitfires continued to climb, their orders being to engage the escort, leaving any bombers to the Hurricane boys. However, for some strange reason, perhaps they were just a decoy, the reported bombers turned away to the south east without dropping any bombs.

The Tigers found the Messerschmitts, initially only six to nine in number, of a total of 36, coming in from the sun and heading towards a group of Hurricanes off Dover. Malan turned with the others in behind these 109s without being seen, leading Red Section into an attack. Closing in he selected one Messerschmitt, giving it five two-second bursts from 250 to 100 yards. The German pilot made no violent evasive action moves save for putting his machine into a gentle right-hand turn and decreasing speed. Malan concluded that the 109's controls had been hit. He then turned in behind a second 109 which had flown past his line of flight, and gave it three deflection bursts as he curved round with it. The 109 went down in a spiral, but Malan had to leave as his ammunition was now expended.

It was a pity he had used all his ammunition, but lucky for the pilot who has now been identified as none other than the *Geschwader Kommodore* of Jagdgeschwader 51, Major Werner Molders. Molders was wounded in the action but managed to fly back across the Channel, making a crash-landing at Wissant; his 109 was a virtual write-off. Three of his pilots were not so lucky: Malan, Kelly and Freeborn each destroyed one, and Stevenson and St John damaged two others. Kelly recalls seeing his 109 fall in flames. It was not all one-sided, for Pilot Officer J.H.R. Young was shot down and killed (P9547) and Tony Mould baled out wounded (P9336), and was whisked away to Dover Military Hospital. At the time, Stephen also claimed one 109 damaged, Pilot Officer H.R. Gunn another destroyed and Paddy Stevenson also claimed one destroyed.

On the last day of the month Dover was once again the scene of action, Me109s strafing the balloon barrage. Several squadrons were sent up after them but only 74 Squadron was engaged. Messerschmitts of JG2, 'Richthofen', came down on Flight Lieutenant Kelly's section as they were still gaining height. Pilot Officer Gunn and Sergeant F.W. Eley, his Numbers Two and Three, were quickly shot down and killed. Kelly proved a tougher opponent, twisting and turning as other 109s systematically shot his Spitfire (R6983) to pieces. Even so he was able to get in a burst at one Messerschmitt whose pilot allowed himself to get ahead of the British pilot. Kelly saw it turn away, smoking badly, but then his own machine was hit again. Then the 109s, probably low on fuel, pulled away and left him. His machine was full of holes but he managed to land at Manston without piling up.

On the ground was Jeffrey Quill, Supermarine's Test Pilot, now flying Spitfires with 65 Squadron, who had previously been one of the leading lights in the early development of the Spitfire. He saw Kelly's Spitfire on the ground and was absolutely amazed and fascinated at the amount of damage it had sustained yet it continued to fly on to bring Kelly down safely. H.M. Stephen recalled seeing the armour plate behind Kelly's seat which had a slight mark on one side and a lump the size of an egg on the other, but it had protected his back, thereby saving Kelly's life.

Kelly's 109 was not so fortunate, its wounded pilot crashed at

Fécamp when its smashed engine finally failed.

Another event which occurred on the 31st was the award of a bar to Sailor's DFC, and a DFC went to Johnny Freeborn – both popular awards. Freeborn was a Yorkshireman who had now destroyed five German aeroplanes. Malan's award was for the destruction of the two Heinkels at night, and his own score was now nine destroyed or shared destroyed with another seven damaged or probably destroyed.

As August began so too did the next phase of the Battle of Britain. The July weather had not favoured any mass assaults by the Luftwaffe but as August began the weather improved. It seemed certain that the long awaited attack would now come.

Just in time to help meet this attack came two Polish pilots to the squadron, Flight Lieutenant Stanislaw Brzezina and Flying Officer Henryk Szczesny who arrived from 5 OTU. They were quickly nick-named 'Sneezy' and 'Breezy', this being far easier than trying to pronounce their names. Both were to serve with distinction with the Tigers, Szczesny flying as Malan's Number Two on several occasions. It is highly probable that he and Malan got on so well because they were both about the same age, Malan now 29, Szczesny 31, and both had been flying for some time in their respective air forces.

Szczesny, also known as Henry the Pole, was born in Warsaw, had received a college education, then became an army cadet before joining the Polish Air Force, going to the cadet school at Deblin, Poland's equivalent to Cranwell. He was commissioned in 1933, joining initially the 5th Air Force Regiment then onto the 3rd Regiment as a fighter pilot. He married in 1937 and had two sons. In September 1939 when the Germans invaded his country he flew against them in P7s, P11s and was the only pilot to fly against the Germans on the Eastern border in the P24 'Kobus'. In battle he destroyed two Luftwaffe aircraft and probably two more before being wounded in the left leg.

Following an operation in a Bucharest hospital he escaped on a Greek ship, sailing to Malta, then went to France and finally to England. In February 1940 he found himself at RAF Eastchurch . . .

'. . . to begin to learn the English language. In May posted to RAF Manston with my platoon of Polish Cadets, as CO. Later to Blackpool. In July posted to OTU at Old Sarum near Salisbury and in August for five days to OTU – Spitfire – at Aston Down and on 5th August, I and my friend Flight Lieutenant Stanley Brzezina were posted to RAF Hornchurch to 74 'Tiger' Squadron, where shortly Sailor Malan became Commanding Officer.

'At once we have got our nick-names Sneezy and Breezy and I also Henry the Pole. I was happy as a lamb to be in the air once again and on Spitfire I – superior to Me109, in Tiger Squadron and like Tiger, to kill.'

CHAPTER FIVE

Leading the Tigers

It was a supreme moment when Sailor Malan was given command of Tiger Squadron on 8 August 1940, following Squadron Leader White's posting to Headquarters Fighter Command. It was a natural choice, for Malan had been leading the squadron in the air since May. It was a popular promotion, too, with all of the Tigers. His charisma had captured them all even before the shooting war began and now, with the DFC and bar as proof of his prowess, if proof had been needed, he was truly the 'boss'. From acting pilot officer to squadron leader in less than four years with the same squadron was a crowning achievement.

For Sailor it all had to be taken in his stride, for the squadron was still very much in the front line and the Battle for Britain was about to enter a crucial phase. He would, of course, continue to lead the Tigers from the air, not from the ground. He had already more than proved that he was a fighting leader.

The squadron's first major action under Malan's leadership came on Sunday, 11 August. It started out a fine bright day. The German's first gambit being a series of marauding *frei Jagd* (free chases) by Messerschmitt 109 fighters over and across the Dover area; then twin-engined Me110s of the special Erprobungsgruppe 210 (EG210) dropped bombs over the harbour town. The intention was to draw RAF fighters into the air, for the Luftwaffe's task now was to destroy all the Spitfires and Hurricanes they could in order to gain air superiority over the south of England in preparation for a possible landing of troops in an invasion. Although these first raids of the day were not intercepted, RAF patrols in the south-east had been active and now needed support as they began to land and refuel. Air Vice-Marshal Keith Park, AOC of 11 Group Fighter Command, ordered his Sector Stations to send up fresh patrols. 74, already down and refuelled at Manston, became airborne at 7.45 a.m. scrambled to go after the Me110s. They were too late. Then a new enemy force was reported at 13,000 feet, coming in towards Dover. Malan turned his squadron onto an east-north-easterly course, grabbing as much height as possible, and eventually reaching 20,000 feet high in the sun. He then turned them down sun and towards Dover, surprising a formation of eight Messerschmitt 109s flying at about the same height. The Messerschmitts were flying in pairs in a staggered line astern. Immediately he ordered his Tigers to attack.

The 109 pilots saw them and all immediately put their machines into a quick half-roll, then dived, the German fighter pilot's usual method of getting out of trouble fast. Malan looked behind and above, searching the sky. It seemed clear, so taking a chance he followed the diving 109s, latching onto one, overtaking him 2,000 feet lower down. He commenced firing at 200 yards, the 109 beginning to level out at 12,000 feet, where Malan gave it another two two-second bursts, again from 200 yards. The German pilot again flicked over onto his back and dived towards the distant French coast. Malan stayed with him, closing right in to 100 yards, firing two or three more bursts, then suddenly the Messerschmitt burst into flames and was obscurred by heavy smoke. They were now down to 4,000 feet about one mile to the north-west of Cap Gris Nez. As the 109 went down Malan left it and turned back towards England.

Several 109s had been hit and claimed, 74's only casualty being Paddy Stevenson (P9393) whose machine was crippled, forcing him to bale out. He splashed down into the sea and was fished out safely.

Other 109 formations engaged at this time began to exhaust 11 Group's aircraft availability and this became the pattern of the morning. Seeing the obvious danger, Keith Park did not allow himself to be drawn in total but 32, 54 and 74 Squadrons were sent up again.

Malan led 74 off at one minute to ten, ordered to engage 109s that were approaching Dover yet again. He led the climbing fighters on a north-easterly course to 24,000 feet before sweeping round to the right in order to approach Dover from seaward, thereby hoping to surprise any 109s he found from behind. Below he saw several small groups of 109s about mid-Channel but as 74 approached most of them dived towards the French coast.

Then he spotted two 109s and led Red Section down on them. Coming in fast he fired from 150 yards but his speed was too great, causing him to overshoot. Breaking away, his section continued to attack. Malan regained height to where Blue and Green Sections were still patrolling. He tried to attract their attention but his radio was playing up and he was unable to contact them. He therefore turned towards Dover on his own, still looking for Messerschmitts.

He saw two at 25,000 feet, again at about mid-Channel, attacked and gave the rearmost two two-second bursts, seeing his bullets splattering on its grey fuselage. It immediately flicked over and away to the left, leaving Malan clear to fire at the leading 109. His bullets again found their mark, confirmed by the appearance of a stream of white vapour. Then eight 109s which Malan had failed to see earlier began to dive down upon him, but Malan immediately put his Spitfire into a righthand spiral which they did not follow.

Coming out, now alone again, he spotted ten more 109s. He endeavoured to attack a straggler but was unable to close the range without exposing himself to a turning attack by the leading Messerschmitt. He tried again to raise the rest of the squadron but the atmospheric conditions made the R/T useless, so he broke off the pursuit, landing back at Manston at 10.45.

Completing his second combat report of the day, he claimed two Me109s as damaged, but it was still early. At 12.45 the squadron was off again but without Malan. Yet again the Tigers met 109s and 110s over the convoy 'Booty' off the Essex coast, claiming several but losing two pilots, Pilot Officer D.G. Cobden and Pilot Officer D.N.E. Smith.

Malan led the squadron on its fourth sortie at four minutes to two o'clock with orders to patrol above Manston at 10,000 feet. The convoy was now nearing the Thames Estuary and heading for it came 45 Dornier 17s with a formation of Ju87 Stuka dive-bombers. Above them was the inevitable Me109 escort.

The Spitfires nosed through a layer of 10/10ths cloud; Malan was leading just eight Tigers. Clearing the opaque whiteness, they saw about thirty Ju87s 2,000 feet above and a half a mile behind

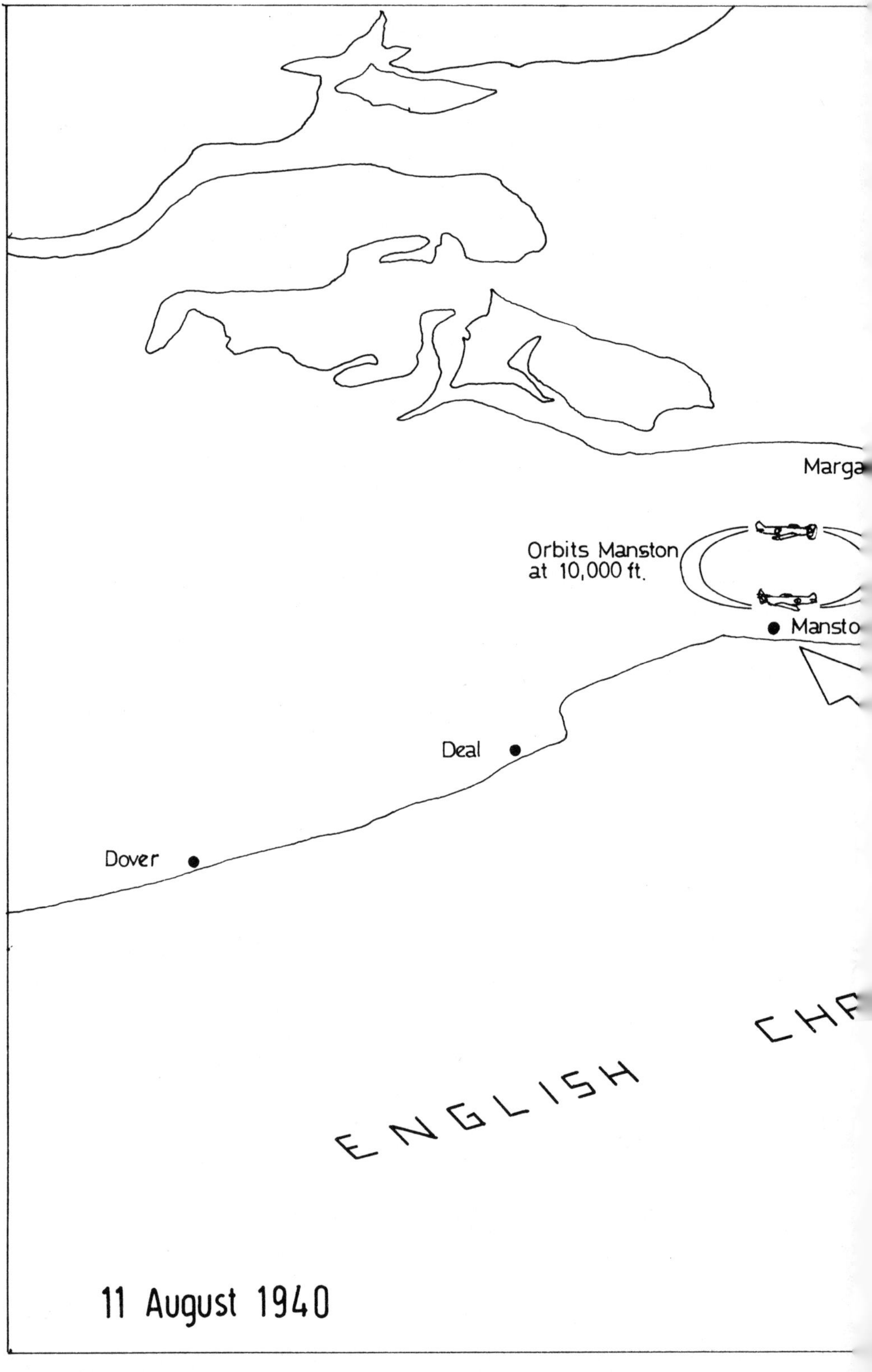

11 August 1940

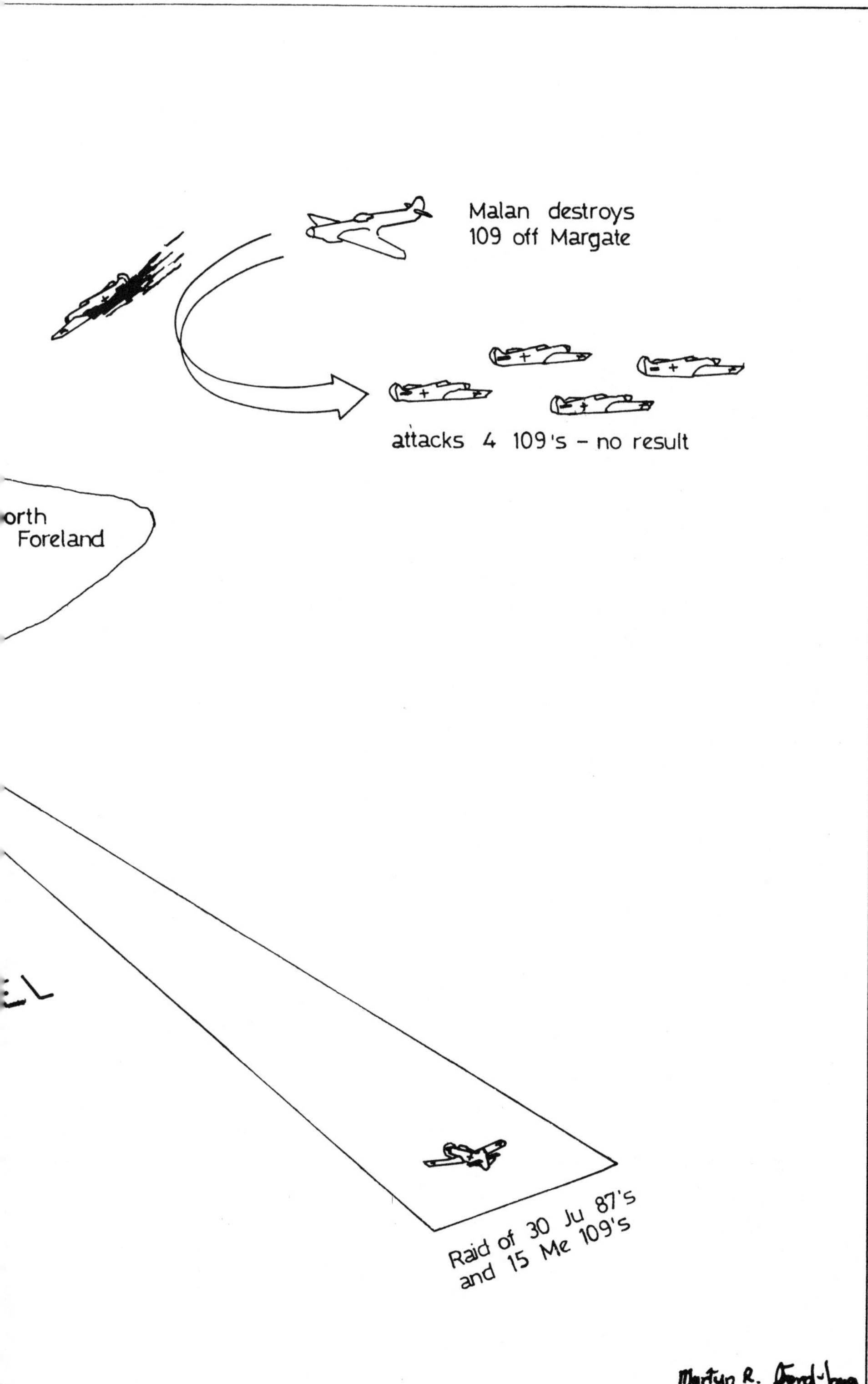
Malan destroys
109 off Margate
attacks 4 109's – no result
orth
Foreland
EL
Raid of 30 Ju 87's
and 15 Me 109's
Martyn R. Ford-Jones

them. The Germans also saw the Spitfires at once and began diving towards a gap in the clouds whilst the Messerschmitts began to close up with them.

Ordering Blue Section to attack the dive-bombers, Malan closed with the Messerschmitts, attacking one as it dived through a gap. He began firing at 200 yards with a 30 degree deflection shot, closing to 100 yards dead astern. After his third burst the 109 burst into flames and dived into the sea off Margate. Malan immediately climbed towards the cloud before diving on another group of four 109s using up the remainder of his ammunition in dispersing them.

The Squadron's attack, brilliantly led by its CO, Sailor Malan, effectively broke up the raiders and saved the convoy from substantial damage. This ended a hard day's fighting. Four patrols had been flown and most pilots had flown on three of them. A number of Germans had been claimed for the loss of three Spitfires and two pilots.

The next day the squadron rested, patched its damaged aeroplanes and prepared for the next action. That action came on 13 August – the German's '*Adler Tag*' – Eagle Day. In order for the Germans to be in any sort of position to invade, the Luftwaffe just had to gain freedom of action for its aircraft. On Eagle Day began the all-out assault against the Royal Air Force, planned to destroy its aeroplanes in the air and its bases on the ground.

Malan and 74 Squadron met the first German raid of the day, scrambling from Hornchurch at 5.55 a.m. They headed out over the Thames Estuary, finding a force of Dornier 17s making for the English coast. They came from KG2, flying in without fighter escort as the bomber leader had failed to receive a recall signal sent out because the weather had deteriorated. Five squadrons of RAF fighters had scrambled to meet the raid but only the Tigers found them.

When the Spitfires saw the unescorted bombers everyone was keen to get in and have a crack at them. Piers Kelly recalls that the Tigers were all 'elbowing' each other out of the way so intent were they to get behind a target.

Malan led a pass on one vic of three Dorniers, and closed in to within 100 yards before raking them with machine-gun fire.

(Above) Sailor with his dog 'Peter'. With him is B V Draper and H M Stephen.

(Left) Piers Kelly as a Wing Commander later in the war.

Pulling round he got in behind the number three bomber and fired again from 150 yards. Hammered with four two-second bursts the Dornier burst into flames in mid-air and headed towards the water.

The leading Dornier now received Malan's attention as he gave it three two-second bursts from 150 yards. One of the bomber's engines took hits, shed bits and pieces and the propeller jerked to a stop. As Malan broke away he felt confident that the Dornier would be unable to reach its base. He then finished his ammunition on Dornier number three.

Piers Kelly, the B Flight leader, had also attacked a vic of Dorniers, his fire forcing out one of the formation trailing smoke. He too then tackled a second Dornier, fired, but was then hit by the gunner from the third Dornier, and a bullet punctured his glycol tank. Immediately his cockpit was filled with fumes and a white/grey haze. He pulled up and away from the bombers, unable to see outside the cockpit. His first thought was to bale out but Malan, ever watchful even in the heat of action, saw Kelly's predicament and called on the R/T:

'Come on, Kelly, follow me!'

Malan formated onto Kelly's wing, giving him instructions, got him back to Manston, and even guiding him down to land on a good strip of airfield not pitted with a crop of new bomb craters. Kelly was tremendously impressed by the sheer presence and determination of Sailor Malan. In the air he always seemed to him to be watching over everyone as well as taking tactical control of the battle. Piers Kelly certainly was thankful for Malan's skill and watchfulness on that day.

It was while operating from Manston, 'Hell's Corner', which was in a first class position to be strafed and bombed by marauding German fighters and bombers almost at will, that an incident occurred which H.M. Stephen remembers.

As 74 Squadron landed fresh from battle, and Manston had quite obviously been attacked yet again, a detachment of 74's fitters and riggers who were, of course, at Manston, met the Spitfires and in their usual efficient way set about preparing them for action as quickly as possible. However, the station armourers did not appear and a Spitfire without ammunition was useless.

Malan was furious and Stephen recalls that when Malan was

eventually roused to anger it was fearsome. He grabbed a belt of ammunition, raced into a nearby air-raid shelter and hit the first armourer he found with it, yelling for them all to get out and rearm his squadron. He was going to help win this war and by God they would do their bit.

*

It was now time for a rest. It had been important that Manston was manned by the RAF, if for nothing else but to show some tangible morale for the local population. Yet the squadrons using it were, more than any other, right up in the sharp end. Not unnaturally these units had to have a break and on the 14th, 74 Squadron flew to Wittering from their parent station at Hornchurch, for a breather. Malan was aware that his men could use a breather but he was, nevertheless, angry at being pulled away from the south where it was all happening. Almost immediately he was agitating to be allowed to get at the enemy again as quickly as could be managed.

A week later 74 moved north to Kirton-in-Lindsay in Lincolnshire where four new pilots arrived to boost its ranks. In from Number 7 OTU came Pilot Officers E.W.G. Churches and R.L. Spurdle, a New Zealander, and from 142 Squadron came Flying Officers W.D.K. Franklin and R.J.E. Boulding. Roger Boulding, who eventually retired from the RAF as a wing commander, says:

> I flew with Sailor on many occasions, sometimes as his Number Two. I do have vivid memories of him as a leader, tactician and commander and believe his influence on the tactics adopted in the later stages of the Battle of Britain and the subsequent offensive action over France was considerable. In the period I knew him I believe he was just about the best we had. He wasn't easy – but we all respected him greatly.
>
> I was posted to 74 Squadron at the end of August 1940, having been with the Advanced Air Striking Force in France flying Fairey Battles. 74 moved from Hornchurch at the same time to Kirton and shortly after, to Coltishall. Sailor was the squadron commander and whilst we were at Coltishall the

other resident squadron (Hurricanes) was commanded by Douglas Bader – which made an interesting contrast.

Piers Kelly left to become an instructor, and B Flight was taken over by Mungo Park.[1] Johnny Freeborn was promoted to flight lieutenant and confirmed as A Flight commander. Then on 3 September, the war's first anniversary, Malan and Freeborn went to Buckingham Palace to receive their decorations.

Three days later Malan and Stephen went to 12 Group Headquarters to have an interview with the AOC, Air Vice Marshal Trafford Leigh-Mallory. This was at the time that Leigh-Mallory was strongly advocating the use of a Big Wing of two or even three to five squadrons. The object was simply to hit the Germans en-masse, rather than nibble at them at just squadron strength, and be constantly out-numbered. The argument against this plan was that by the time such a large fighter force could be assembled, the Germans would have attacked and be on their way home. Malan and Stephen gave their views, and Malan stated strongly his own thoughts on attacking the enemy. As Stephen recalls, Malan was happy enough to be part of the Big Wing idea, if only as a way to get back into action. The only thing he insisted upon was that he and his Tigers should be on top in the air, above everyone else. Even at this early stage he had learned that to be above was to control the action.

Four days later, on 10 September, 74 moved to Coltishall in 12 Group, where it would fly down to Duxford and operate with the Big Wing – The Duxford Wing.

The chance for action came on the 11th. The Germans had started attacking London, giving the more southerly RAF fighter bases the respite they desperately needed. They had been bombed and strafed for most of August and were near to collapse. With the switch to London the Germans had farther to travel, thus allowing the 12 Group Wing time to form up.

On the 11th, a Wednesday, it started out wet but by the afternoon the weather had improved. By 3 p.m., the radar screens began to show German bombers massing over the Pas de Calais

[1]Piers Kelly later commanded a night-fighter squadron in North Africa, where he won the DFC.

and, soon afterwards, being joined by more than 200 fighters, and proceeding across the Channel, heading for the Thames Estuary. They began to be harassed by RAF fighters but little impression was made. Then they reached the outskirts of London where most of their escort was forced to leave them as fuel began to run low, but they were replaced by others.

The Tigers with Malan at their head, flew as rear-guard of three squadrons of the wing, led by 19 Squadron (which was in fact a composite of aircraft of 19 and 249 Squadrons), with 611 Squadron in the middle. They were required to engage the bombers, now over London itself, at 20,000 feet. It had been agreed that 19 and 611 should engage any fighters while Malan's unit went for the bombers.

At approximately 4.30 p.m. the wing saw a long rectangle of Ju88s at 20,000 feet. 74 was flying in three sections of four in line astern. Malan gave the order to attack from head-on but before they could get into position they saw enemy fighters (identified as He113, but they were, of course, Me109s) coming down onto them. Malan was forced to abandon his plan and turned immediately towards the bombers so as to deliver at least some form of attack before the fighters engaged them. Unable to turn completely in the space ahead of him, Malan quite accidentally managed a very effective beam attack at close range by turning right across the front of the bombers. Flying across them he raked the formation from 150 to 50 yards seeing his own fire chew into two Junkers in the tightly packed formation.

The German fighters were now very close, so Malan continued down in a fast spiral to 13,000 feet, then he pulled up, climbing to 20,000 feet. He then saw two bomber formations at 22,000 feet, one of 30, the other of 50. Malan headed towards one of them, but as he closed, saw a lone Ju88 2,000 feet below, which he thought was either experiencing engine trouble or was observing the results of the other bombers' attack. Malan came screaming down in a head-on attack, blasting large pieces from the 88's port engine. Pulling round in a tight turn he attacked again from 100 yards, just as a Messerschmitt dived down at him. Malan continued down, but looking back, saw the Junkers falling away in a gentle dive, its port engine well alight. As he watched he thought he saw a parachute open close by the bomber. The German fighter, which he again thought was a Heinkel 113, followed him down to 10,000 feet where he pulled his Spitfire (X4068) into a steep, blood-draining left hand turn, blacking out in the process, but it effectively shook off the German.

Johnny Freeborn, leading Yellow Section, was unable to get to the bombers before the fighters intervened. He headed southwards and made a head-on attack upon a Dornier following this up with three stern attacks. The Dornier burst into flames and crashed near Dungeness, where he noticed a wrecked Heinkel 111 in the adjoining field.

Everyone returned safely, some landing at Duxford, some at Coltishall.

The squadron began to re-equip with Spitfire Mark II. At this time, Roger Boulding recalls:

> We re-equipped with Mark II Spitfires at Coltishall and I recall two incidents during this period which reflect in some degree the personality of our boss. In one, myself and two others who had also been transferred from Fairey Battles and were equally new to Spitfires and fighter tactics, were on patrol over East Anglia when we intercepted an enemy aircraft (Dornier I think). We three chased it with great enthusiasm and not too much skill. The German crew were obviously experienced and made good use of the available cloud cover and various defensive armaments so that in the end we lost him. On our return to Coltishall, Sailor gave us a pretty cool reception – his squadron had lost what should have been a certain kill. At that time the standard setting for one's gunsight was a range of, say, 350 yards. Sailor's orders were to set them at 250 yards – to make sure we got close enough! He had me so petrified that when he asked what range I had my sight set at I gave the wrong answer (350 yards) and I was never sure he quite accepted my subsequent assurance that I had in fact set 250 yards!
>
> The other incident concerned our much appreciated Mark II Spitfires. Whilst the squadron flew from other fields further

> south in a daily reinforcement role those with less experience were left to air test the new aircraft at Coltishall. On one such trip (12th September) I found it impossible to lower the undercarriage and in spite of a whole range of manoeuvres eventually had to make a wheels-up landing. As this was in the days before we had concrete runways, I put it down as gently as possible in the middle of the field – in fact, as it happens, with minimal damage. The first man to reach the aircraft as I clambered out was Sailor who, having quickly established that everything was switched off etc, pointed out that I could have saved the groundcrew a good deal of time and effort if I had put the aircraft down alongside the hangar instead of the middle of the field! He was right of course, but to a young pilot, very inexperienced in Spitfires who had just had a rather hairy experience it was a slightly startling reaction! For some time afterwards I was worried that in some inexplicable way it had been my fault the undercarriage didn't come down and that Sailor suspected I had boobed so I was most relieved when some days later the squadron engineer officer reported that there had been an error in the assembly of the hydraulic control box.

Changes took place over the next few days, Paddy Stevenson going to 5 OTU at Aston Down as an instructor, but receiving the DFC. Stanislaw Brzezina went to take command of the newly formed 308 Polish Squadron and Warrant Officer Ernie Mayne, who had flown with the Royal Flying Corps in World War One and who had flown gallantly with 74 Squadron for five years, left with Kelly to go to 6 OTU as an instructor. He was just on forty years of age, but had one or two scalps to his credit before age and the Air Ministry caught up with him.

Although the squadron had combats on the 14th, 23rd and 24th, Malan was not among the scorers. Sergeant David Ayers was lost on the 23rd; he was seen to bale out over the sea (P7362) but was not found. His body was washed ashore on 4 October. The squadron was left out of the final September battles but Steve Stephen, leading Green Section off Yarmouth, chased and damaged a Heinkel on 1 October.

At Readiness. l. to r.: Roger Boulding and 'Sam', Henry Szczesny, Johnny Freeborn and Steve Stephen.

Biggin Hill October 1940. l. to r.: Mungo-Park, Johnny Freeborn, Ben Draper, W D K Franklin and Peter StJohn *(front right).*

The squadron was still training new pilots. Four had arrived at the end of September, but one of them, Pilot Officer Buckland, was tragically killed together with veteran Pilot Officer D. Hastings following a collision in the air. However, a change was in the air, and on 15 October, 74 moved south again, to the RAF's premier station – RAF Biggin Hill. Almost before they had started to unpack they were in action, Pilot Officer Nelson and St John each claiming an Me109.

One of the controllers at Biggin Hill was Flight Lieutenant W.A.K. 'Bill' Igoe who was to come to know Malan fairly well in the mid-war years. His first encounter with the South African, however, was something very different.

> My first contact was when he was CO of 74 Squadron. He called me up asking permission to land on his squadron, upon being posted to Biggin Hill. I told him to land individually. He told me his squadron always landed as a squadron. I told him he did what I ordered. The aerodrome had just been bombed and was full of craters; he could not know that and I could not tell him in clear language for obvious reasons. Ten minutes later the duty pilot told me that they had landed in squadron formation, whereupon I threatened him with all sorts of dire consequences and told him Biggin Hill squadrons always obeyed orders. Anyway, from an explosive beginning we became very good friends.

Roger Boulding continues with his story:

'We moved to Biggin Hill on 15 October. By this time the main activity over the 11 Group area consisted of sweeps by Me109s in the fighter and fighter-bomber roles with only occasional formations of bombers on a relatively small scale. The recurrent problem for us was that we usually had to climb up from our bases with the enemy well above us – always a disadvantage and most uncomfortable. A considerable number of our squadrons were still flying formations made up of basic elements of three aircraft in vic formation – usually with one aircraft over the squadron (called a weaver) whose job it was to watch the formation's rear. This tactic was comparatively unwieldy and furthermore the "weaver", by the nature of his job, used more fuel than the rest of the formation which made him very vulnerable. Sailor's answer to this problem was to fly the squadron in three sections, each of four aircraft in close line astern, with each section sub-divided into two. The utmost effort was to be devoted to staying together in pairs so that one could concentrate on the enemy whilst the other guarded his tail. Using this formation the whole squadron could be manoeuvred in a far more flexible fashion and maintaining close formation in line astern was a far easier proposition – so enabling all pilots to give more attention to looking for the opposition. Instead of one flying as a weaver, the whole squadron could weave as one. My recollection is that this was one tactic that Sailor played a major part in perfecting – if he didn't actually invent it himself.

'Subsequently, when we began to go over to the offensive on sweeps across the Channel, this formation was adopted to the reverse situation when we were at altitude and the enemy were obliged to climb up to meet us. Under those circumstances we needed to be able to cover the maximum possible space so that we could spot the enemy at the earliest moment to enable us to make full use of our height advantage. So Sailor had us open out the basic sections of four aircraft so that there was a considerable gap between each pair in the section. The pairs could then weave semi-independently, the squadron maintained contact as a unit which could still be manoeuvred quickly and we covered a much larger piece of the sky. This formation was later abandoned in favour of the "finger four" type, but that was after I was taken PoW.

'Of the sorties I flew as Number Two to Sailor, one of the earliest after we arrived at Biggin Hill is still vivid in my memory. We were still climbing above the airfield area attempting to intercept a fairly large formation and, from the ground controller's comments, were apparently quite close when two Me109s crossed in front of us in a shallow dive from left to right. Any enthusiastic pilot could have been expected to have turned onto the tail of these two sitting targets and followed them down but Sailor knew his stuff, resisted the temptation and almost immediately spotted a large formation positioning themselves so that they would have been able to jump us from an ideal vantage point had we fallen for the bait. There ensued a terrific mix up with the opposition, and mine was the typical experience of those new to this particular game. One moment the sky is so full of aircraft you don't know quite what's going on and then with two or three minutes you are on your own with nothing in sight at all! I'm ashamed to say I lost Sailor in all this tho' I should have hung on to him to guard his tail. Did he take me as his Number Two so I could learn?'

Number 74 Squadron was now fully equipped with the Mark II Spitfire and all hoped for better luck than of late. However, the Battle had now entered its fourth and final stage. The massed raids by German bombers gave way to odd nuisance raids or penetrations by bomb-carrying Me109s, which having dropped their bombs, became freelance fighters. It was a time which had its own particular brand of danger for the RAF fighter pilots. The 109s were always above the climbing British fighters and in the hazy, cloudy October and November days, death lurked ever ready to claw down the unwary.

Another contemporary of Sailor's at this time was Flight Lieutenant R.M.B. Duke-Woolley. Duke-Woolley had shot down a Heinkel on the same night as Sailor shot down his first two, being so nearly the first RAF fighter pilot to bring down a night raider.

'I remember a misty morning at Kenley in early October 1940. A day of ground mist which would later "burn-off" as the sun hotted up. I was temporarily in charge of 253 Squadron as a flight lieutenant and we had landed the previous evening at Middle Wallop. The pilots all wanted to get back to Kenley and their razors and tooth brushes; the weather was fine in Hampshire and we could obviously return if necessary. So we set off for Kenley where visibility was reported poor.

'I spotted the airfield after a bit of a search and visibility was very poor indeed laterally, and not very good vertically because of a low sun shining through the mist. However, by forming everyone into a long line astern in orbit roughly over the field, then leaving them to land myself, then immediately taking off again and going into an orbit off the runway, and finally having the long line reform onto me, we were able to reverse the whole process by leading off downhill in orbit and landing one by one on the runway. I thought the whole performance was rather neat with everyone keeping sight of the one ahead.

'Arriving at dispersal I found it all cluttered up with Spitfires, which was Sailor and the mighty 74 Squadron. Sailor, who had watched us land, said nobody was going to catch him "taking off in this stuff." He rather nicely said that his squadron could not possibly have landed in the conditions and congratulated me on the squadron's performance.

'I knew Sailor off and on for some years but never all that well. He was a very "nice" man, not at all the swashbuckler which one or two other aces could well be called. He was at Biggin Hill, of course, in 1940 while I was at Kenley. He was about average at Staff College, tending to be too modest to thump the table – as others might have done with his reputation. I think he was a hunter, really. Good shot, indifferent to danger and with the same killing instinct as a professional hunter,'

*

On the afternoon of 17 October, Sailor Malan led his squadron away from Biggin Hill to intercept fighter-bomber raids approaching London. He led them up-sun to 26,000 feet turning towards some bursting AA shells over the Thames Estuary.

At 3.30 p.m. they suddenly saw two yellow nosed 109s crossing ahead of them and they made an attack on them from out of the sun. Malan closed with the right-hand Messerschmitt giving it two short bursts with a quarter deflection from 200, down to 150 yards; then he let go another two bursts which smashed into the German's elevator, probably hitting the control wires, for the 109's nose went vertically downwards very suddenly, instead of the pilot going into the more usual half-roll and dive.

Malan attempted to follow, but the Spitfires of the period did not have the Messerschmitt's fuel injection system, so Malan's carburetter was starved for several moments, causing his engine to stop momentarily. When it picked up again, he closed to 150 yards on a half-roll where he gave the 109 another four second burst. He then found himself flying an aileron turn in order to keep direction; then he let go another longish burst. The 109 began to leave a trail of smoke, but then Malan blacked out, completely losing consciousness for a couple of seconds. He lost sight of the 109 and eventually pulled out at a height of 9,000 feet, above a layer of 10/10th cloud. When he landed he found that only his four starboard guns had fired.

Others had also become engaged with 109s, three being claimed as destroyed although the Tigers lost Flying Officer Alan Ricalton (P7360) who was shot down and killed by 109s in that dangerous autumn sky.

On the 20th more 109s were engaged. Mungo-Park and Stephen each claimed one, Draper probably another, but they lost Sergeant T.B. Kirk. He had damaged a 109 when his Spitfire was hit over Maidstone and he was wounded. He died later in hospital. Sergeant C.G. Hilken was also shot down, took to his parachute and was taken to hospital with wounds.

The following Tuesday, the 22nd, Malan led nine Spitfires, being given vectors to intercept single high raiders but made no contact. Then they were ordered to join 92 Squadron who were patrolling at 30,000 feet over Maidstone so as to intercept raids now coming in from the south. They stayed at 30,000 feet in order to push out exhaust condensation trails with the object of trying to frighten the Messerschmitt pilots.

Flying over Ashford, 92 Squadron saw enemy aircraft and gave a 'Tally-Ho' but Malan, flying in a circle, could see nothing. Then at 2.10 p.m. he suddenly saw six Me109s below at 26,000 feet, flying on a south-easterly course. Having become separated from four of his men (Yellow Leader had dropped out with oxygen trouble, his Number Two following, while Yellow Three and Red Two had been unable to keep up) Malan winged over and dived with the remaining four. Malan attacked the leading 109 in a fast dive, thumbing the gun-button at 200 yards. It started to smoke heavily after his second burst but the German pilot levelled out at 8,000 feet. Malan continued to fire at it but it did not go down and he had suddenly to break away as an ice layer formed on his armoured windscreen.

Still watching the crippled 109, he saw it cross the coast near

Squadron Leader Sailor Malan DSO DFC & bar with the statuette presented by the Dutch East Indies.

Sailor with his pilots at Biggin Hill, October 1940 (note side arms). l. to r.: Sergeant Kirk, W D K Franklin, A L Ricalton (?), (?), (?), Roger Boulding. Henry Szczesny, Malan, (?), Mungo-Park, Ben Draper, (?), Peter StJohn, (?), (?), Pilot Officer P Chesters. Kirk, Ricalton and StJohn were all killed during October.

Hastings. Malan remained above, still looking at the smoke trail left by the Messerschmitt. Then about five miles out from the coast the trail went down and he saw a splash. This was Leutnant Kurt Muller of 3/JG51 who got out of his machine to begin the swim to shore. He was rescued and taken prisoner. The squadron claimed a second 109 but these kills did not come without losses. Two Spitfires did not get back; Pilot Officer P.C.B. St John was killed (P7341) and Pilot Officer R.G. Spurdle parachuting from his crippled aeroplane. Peter St John's loss was felt keenly by the squadron, for he had an infectious, happy disposition, always laughing and full of fun.

The battles continued in the autumn skies and several 109s were claimed during the remainder of October. Then more newcomers arrived. Among them was Pilot Officer W. Armstrong, posted in from 54 Squadron. Then Flying Officer Nelson was killed in action on 1 November, just as the Battle of Britain officially ended. His was another sad loss. The very next day Malan led 74 and 92 Squadrons against more 109s, Mungo-Park forcing one to fly off trailing smoke and coolant, while Pilot Officer Churches shot another into the sea. Sergeant Skinner got another in flames while 'Kiwi' Spurdle put one into a field outside Ashford. On the 5th Mungo-Park received the DFC, and H.M. Stephen was awarded a bar to his DFC.

As the winter weeks began, the air actions gradually occurred less and less. Sailor Malan, who had been leading the Tigers in the air since May, took some well earned leave, only to miss a great party. Mungo-Park led the squadron on 14th November, in company with 66 Squadron, finding about 50 Stukas, plus 25 Me109s. 74 claimed a total of 15 destroyed. When Malan heard the news he sent his lads a telegram: 'Congratulations you rats – Sailor.'

Malan scored his next victory on 23 November shortly after 12.30 p.m. Leading 12 Spitfires and in company with 92 Squadron, they patrolled above Maidstone at 25,000 feet, then they were ordered to sweep over Dover. There they found several groups of Me109s who appeared reluctant to fight. Later they found two 109s at 27,000 feet, midway between Dover and Cap Gris Nez. He climbed 500 feet to engage, attacking with short bursts with half deflection. The 109 began to leave a long trail of white smoke and it then began to fly in a wide circle before diving towards France, with Malan following. The 109 started to draw away but Malan opened up at extreme range, then slowly closed the gap to 150 yards, then closer to 50 yards, before emptying his guns into it. The 109's cockpit canopy tore back as black smoke replaced the white. Then it caught fire right over Gris Nez itself.

Four days later, on the 27th, Malan and the squadron were flying on patrol at 25,000 feet above the Isle of Sheppey, when he was attracted by AA fire. Investigating, he spotted two Me109s flying south about 500 feet above. Turning the squadron he gave chase, but the Germans saw them and over the radio they heard a German voice yell *'Achtung, Spitfeur!'*

The 109s began to dive steeply towards a haze level but pulled out at 18,000 feet, allowing Malan to close to 800 yards. The two Messerschmitts were still descending, flying 200 yards apart. Closing in still further Malan began firing at 300 yards but the German pilot, seeing tracer shells zipping over his head, broke away rapidly. For a moment Malan lost sight of the 109s but picked them out again, at 5,000 feet over Dungeness.

Closing in again, to 150 yards, he fired as he got in closer and closer until he was down to 50 yards. Then the Messerschmitt simply blew-up and disintegrated; the wreckage tumbled down into the sea, thirty miles out from England. The second 109, also hit, was smoking badly. Malan recorded: 'To judge by its gyrations either the machine was badly damaged or the pilot was badly frightened.'

Flying with the squadron was a Squadron Leader H.J. Wilson, on detachment from the RAE, Farnborough, gaining combat experience. This seemed a good opportunity to evaluate properly an air action, so Malan ordered Hugh Wilson in to finish off the 109. He did so, and the 109 went into the sea.(1)

With 74, 66 and 92 Squadrons now firmly ensconsed at Biggin Hill, each chopping away at the elusive Messerschmitts, it was realised on the Station that since the war began, the total victories

(1)Later Group Captain H.J. Wilson CBE AFC & 2 bars. Chief Test Pilot RAE 1934-37, 1938-44.

claimed by pilots operating from the Biggin Sector was nearing 600. The 500 mark had been passed in the hectic days of the Battle before it had even been realised. Now as 29 November ended, the total stood at 599.

Bets on who would get the 600th and when were placed and a sweep-stake had collected a sizeable kitty for the pilot who scored it.

The following day did not seem too promising. Low cloud and hazy fog kept most of the pilots in bed, glad for the respite. However, John Mungo-Park was short of funds, as H.M. Stephen recalls it, and he was determined to get the prize if at all possible. Mungo-Park and Stephen had been paired together for some time – a result of Malan's scheme to put together those chaps who seemed to get on well together to form pairs. Yet on that November morning Steve Stephen did not share Mungo-Park's keen enthusiasm as he was nursing a bit of a hangover. But he finally agreed to go down to the flights; anything for a quiet life!

His plan of co-operation went wrong as soon as it was reported that enemy aircraft were up and about. Mungo-Park volunteered to fly up through the muck and have a crack, so in company with his overhung wingman, off he went. The roar of their two aero engines brought Biggin Hill to life. The Station Commander enquired who had been mad enough to take off in the foul weather and was told.

'Of all the bloody cheek!' was his immediate comment.

Malan led a convoy of cars to the Operations Room where everyone followed the progress of the chase and subsequent interception.

The two Tigers broke through the mist to find a bright clear sky. They also found eight Me109s, cut one out and then took it in turns to blast it into the ground near Dungeness. The German pilot was wounded but died later. Biggin Hill had its 600th kill, and in sharing the prize with Stephen, Mungo-Park had at least got some of the winnings.

Shortly after mid-day on 2 December, Malan got yet another Messerschmitt. 74 followed 66 Squadron on a patrol over Maidstone at 15,000 feet but when enemy aircraft were reported, they climbed higher towards the Dover/Dungeness area. Shortly afterwards, with 66 at 24,000 feet and 74 at 27,000, the former squadron turned north. Malan leading the Tigers south, spotted groups of Me109s flying to the north.

As these 109s continued past in a fast dive, 74 met them almost head-on. Malan ordered Yellow Section to tackle the leading groups while he turned with Red Section onto the rear bunch. Screaming round and down, Malan got behind one Messerschmitt which turned south-east, into the sun and going flat-out Malan steadied his Spitfire (P7542) and began snapping out bursts at 300 down to 100 yards. The 109 started to leave a trail of heavy smoke as it flattened out at 1,000 feet and its pilot began to pull up into the sun, but it was losing speed all the while. At 5,000 feet Malan gave it a long burst from 50 yards and then his Spitfire was showered in oil as the Messerschmitt dropped into a gentle dive. As Malan went down with it he saw the cockpit canopy blast off and fly back to his right, followed by what he took to be a dinghy pack. They were now down low, ten to fifteen miles out from Dungeness, with Malan still following the obviously crippled fighter. Then it slowly turned onto its back and went into the sea upside-down.

Henry Szczeney left the Tigers on 12 December, going to Stanford-Tuck's 257 Squadron, and later moving to Bader's 242 Squadron before going to 302 Polish Squadron. Later still he went to 317 Polish Squadron. He had enjoyed his period under Malan and remembers:

'I am very proud I was his Number Two on many, many occasions and defending his tail, because he told me to do and to follow blindly his orders. He was shooting down Jerries and I was very close to him – defending his tail. Order is Order.

Once he told me by radio to put Pipsqueak(1) on. Of course, I could not hear him; my radio was always (switched) off – my English was practically nil, so why bother to listen to it? So, poor Operation Room at Biggin Hill, intercepted our 74 Squadron as bandits 12 plus, with Hurricane squadron over Kent. Then Sailor

(1)Identification signal.

show me two fingers up, so I did also show him my two fingers up. He laughed and laughed – after pancake.[2]

'He also introduced me to Winston Churchill, when he was passing to Chartwell, near Westerham, Kent. At Biggin Hill, at dispersal of B Flight, when we came to readiness. He smiled and shook hands with me and asked in his usual deep voice, "Henry the Pole, how many today?" I replied shyly, "Only one Me109, Sir." He said, "Good, many more to come." Then in my broken, pidgin English, standing at attention, saluting in Polish way, two fingers closed together, said, "Sir, please remember Poland was, is and will forever be the Bastion and the wall of Christianity of Western Europe, so make her great, free and independent." My flight commander, Mungo-Park, translated to Mr Churchill in perfect English. Churchill smiled at me again, puffed on his long cigar, shook hands with me once more and bubbled, "We will see on Victory day," and showed me his 'V' sign and drove off.

'Sailor did not want me to leave, and fought like Tiger to keep me in his squadron, offering me B Flight commander because my English was improving rapidly, but no luck – I had to go. (Later) at RAF Acklington I was decorated by AOC 12 Group with Distinguished Flying Cross for all my battles in the air with 74 Tiger Squadron under Sailor Malan. He wrote about me, "Henry the Pole – pilot second to none and so on . . ." Motto of our Tiger Squadron was, "I Fear No Man", but I did add, "Only God and Women."

'Sailor Malan was and always will be in my eyes and in my heart, the greatest pilot and ace number one of the Second World War.'

With his English improving, but far from perfect, one or two of Szczesny's phrases caused a good deal of mirth on the squadron. One was:

'Henry, what is day to-day?'

'Oh boy, today is yesterday!'

And when he got his first English passport, they said he was now a British subject, but Henry replied:

'Oh no, brother, I am British object!'

(2) Order to land, and having landed.

The Orde drawing of Sailor Malan, 29 December 1940.

Sketch of Henry Szczesny VM DFC KW, when a Squadron Leader, later in the war.

On 4 April 1943, as wing leader of four Polish fighter squadrons operating from Northolt and Heston, Henry Szczesny was leading an escort mission to American heavy bombers. The target had been the Renault factory near Paris. Focke Wulf 190s attacked and Szczesny shot down one, then collided with another. He baled out to become a prisoner, ending up in Stalag Luft III, helping with the tunnel used for the Great Escape.

When in prison camp, Szczesny wrote to his former CO, Sailor Malan, warning him to be careful, as: 'All Jerries are after your blood. They already got Bader and Bob Tuck.'

The day following Szczesny's posting, the 13th, some extremely good news was received at Biggin Hill. Malan had been awarded the Distinguished Service Order, H.M. Stephen received the first immediate award of the DSO 'in the field', while Draper and Skinner received the DFC and DFM respectively.

For Malan it came as a fitting end to the fateful year of 1940. In seven months he had more than proved himself in action and led his by now famous Tigers to fresh glories and achievements in this war. A few days later just before Christmas, Malan received a tiny Javanese statuette as a present from the Dutch East Indies, sent via the British Consul General in Batavia.

There was also, for the first time, a list published of the scores achieved by fighter pilots of the RAF. While Air Ministry did not officially approve any form of general publicity for individuals, or so called 'ace lists', a list was required by the public as some form of tangible evidence that a good many RAF air fighters were really damaging the Germans. Fifth on the list was in fact H.M. Stephen with 20 combat victories, two more than Malan, who was seventh on the list with 18. In addition, HQ 11 Group in Fighter Command issued a list of kills for the month of November 1940, and 74 Squadron was top with 26 victories.

Three days before Christmas, the two artists Eric Kennington and Captain Cuthbert Orde, both commissioned by Air Ministry to draw a number of the RAF's top airmen, arrived at Biggin Hill. Amongst those they drew were Sailor Malan, John Mungo-Park and H.M. Stephen – three veritable Sky Tigers.

CHAPTER SIX

Trailing the Coat

What became known as the Battle of Britain had been won. The German Luftwaffe's daylight operations were virtually at an end by Christmas 1940 and to many of Britain's fighter pilots it seemed certain that there would be no continuation by the Germans in the spring. They seemed to know that they had given the Germans a severe drubbing. The surviving defenders were tired and knew they had been in a fight, but they had won. The one thing they wanted to do now was to 'dish some out'.

Sailor Malan's Christmas was disturbed on Christmas Day itself when he, Johnny Kent and Athol Forbes, Commanding Officers of 92 and 66 Squadrons, were summoned to HQ Hornchurch to attend a conference being held by the AOC, Leigh-Mallory, late of 12 Group, now AOC 11 Group. The plan was to fly an offensive operation, hoping to catch the Germans in a seasonal mood. The target for a handful of bombers was installations in the Foret de Guines, and to knock down any Luftwaffe fighters that came up after them. A simple plan if it worked, but it was one that would set the pattern of offensive operations for most of 1941.

In the event the weather produced a thick fog, so the operation was scrapped but it seemed certain that an offensive against Northern France would start when the weather improved. Fighter Command, together with the medium and light bombers of 2 Group, would soon 'trail their coat' on the other side of the English Channel in order to provoke Luftwaffe fighters into action. It would be, but on a grander scale, a similar position to that which the Royal Flying Corps and RAF took in the Great War, when they continually took the air war to the enemy across the trenches and no-man's-land. Now, twenty years later, the Channel had become the front lines.

The abortive Boxing Day operation eventually took place on 10

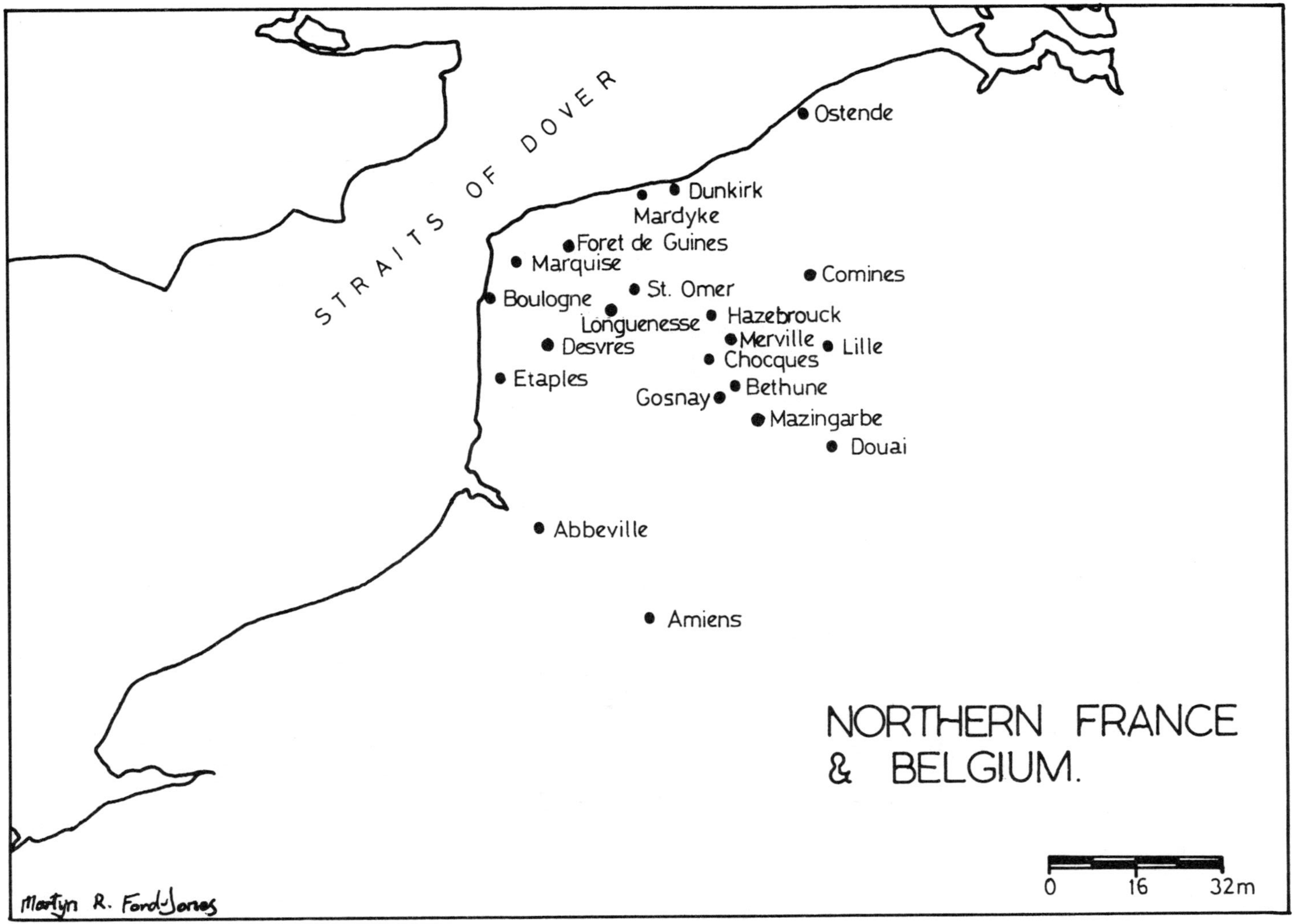
STRAITS OF DOVER
Ostende
Dunkirk
Mardyke
Foret de Guines
Marquise
Comines
St. Omer
Boulogne
Longuenesse
Hazebrouck
Desvres
Merville
Lille
Chocques
Etaples
Bethune
Gosnay
Mazingarbe
Douai
Abbeville
Amiens
NORTHERN FRANCE
& BELGIUM.
0
16
32m
Martyn R. Ford-Jones

January. It was an historic event and Malan, with 74 Squadron, was a part of it.

A morning haze over the Channel prevented an early start that Friday and it was not until ten minutes to noon that the operation began. From North Weald, commanded by the legendary Victor Beamish who led his three Hurricane Squadrons, came 56, 242 and 249 Squadrons. They made rendezvous over Southend with six Bristol Blenheims and a further three Spitfire squadrons, 41, 64 and 611. The three Biggin Hill units were providing what was later to be known as withdrawal cover, sweeping the Channel off the French coast in order to support the raiders upon their return.

Again the target was the Forêt de Guines. The Blenheims turned for the target from the sun after they crossed the enemy coast at Gravelines. German flak groped ineffectually and the close escort Hurricane pilots saw the bombs tumbling down. Then they were away and back over the sea.

A few German fighters made an appearance shortly after the main formation had crossed out of France at St Ingelvert. (Beamish had just shot up some coastal patrol boats). Beamish saw a Hurricane being attacked by what he thought was a Spitfire. The Wing Commander turned to identify the second fighter, then saw it had a yellow nose. Realising it was in fact a Messerschmitt 109, he attacked and shot it down into the sea.

Another pilot in 249, a Polish pilot, Mike Maciejowski, shot down another 109 that he saw landing at the airfield at Guines, then strafed a number of Heinkel bombers on the ground.

However, no excitement came to Biggin Hill's pilots. Johnny Kent led the three squadrons for one hour off the French coast between Calais and Gris Nez areas; 74 flew the rear-guard position. Malan did see some 109s and dived towards them but was unable to engage. The problem, however, was that the Spitfires ran short of petrol. Everyone got down safely, but Sergeant Frieze had to force-land at Detling, was injured in the process and died later in hospital.

The next day H.M. Stephen left the squadron, posted to RAF Turnhouse. At this time he was the top scoring pilot in 74 Squadron. Then on the 13th, news came that Paddy Treacy was safe and back in England. He had escaped and following some incredible adventures got back via Spain. A few days later he visited Malan and the Tigers. It was a great reunion. Unhappily Paddy did not live long to rejoice in his freedom. Given command of Bader's 242 Squadron, he collided with one of his men and crashed into the Channel.

On the afternoon of 2 February, Malan led 74, 92 and 66 Squadrons on another supporting offensive sweep, this time to Boulogne. The main operation, Circus No 2, was against the docks of Boulogne, flown by six Blenheims of 139 Squadron. They were closely escorted by Hurricanes of 601 Squadron. The Biggin Hill Spitfires flew over Dungeness at 1.45 p.m. having gained a height of 15,000 feet, but they continued to climb as they crossed the Channel. At 18,000 feet they approached Boulogne from a wintry sun just below a layer of alto-stratus cloud.

As the wing approached Boulogne, the pilots could see a heavy anti-aircraft barrage and amidst it, the bombers and fighters coming out. The wing flew past Boulogne, heading towards the distinctive feature of Cap Gris Nez. Curving round once more and heading back towards Boulogne, reducing height slightly, Malan spotted German fighters outside the flak zone of Boulogne harbour. He also saw one group of four Messerschmitts at 12,000 feet and led his Red Section down.

Two of the 109s turned towards the Spitfires; Malan guessed the Germans probably thought they were more friendly fighters. He turned steeply onto the tail of one of them and opened up from 100 yards while in a steep turn to the left, using a 30 degree deflection shot of two-seconds.

The Messerschmitt was badly hit and immediately started to smoke heavily and go down. Malan followed in a fast dive and let go three one-second bursts from 150 yards and from dead astern. Oil spewed back from the crippled 109, covering Malan's Spitfire (P7542). Malan broke off at 4,000 feet, but watched the Messerschmitt continue down to dive straight into the outer harbour near what appeared to be a dredger.

Sergeant A.D. Payne claimed a second 109. However, Malan lost his Number Two, Squadron Leader E.J.C. Michelmore (P7741) who had been attached to 74 Squadron since December.

What exactly happened is not known but a few days later it was reported that he had force-landed on a German airfield. 601 Squadron also claimed a 109.

On landing it was found that Malan had expended just 800 rounds from his eight guns. The following day he was interviewed by members of the South African press.

Malan came up against his last German bomber shortly after lunch on 5 February. There was some hostile air activity in evidence, Control requesting 74 Squadron to send up two sections to investigate. Malan led his Red Section in company with Johnny Freeborn's Yellow Section.

Over Dover at 10,000 feet, the Controller radioed to Malan, informing him that a bandit was approaching the English coast between Dover and Folkestone at 4,000 feet. Malan took his Spitfires down through a solid mass of cloud, breaking through at its base of 3,000 feet. He immediately saw a Dornier 215 1,000 feet lower down, turning south off Dover.

Malan ordered the attack. The Dornier made not the slightest attempt to pull up into the cloud. To Malan it seemed that the bomber was much slower than he had met before and that it might have been sent out specially to test new armour-plating. In the subsequent attack the Dornier certainly seemed to be able to sustain a good deal of attacking fire from the rear, and it would go some way to explain why the Dornier did not lift into the cloud.

Malan approached the bomber from below, closing to 150 yards on a zig-zag course, closing to 150 yards before firing. The German rear-gunner opened fire at 400 yards and his Spitfire (P7623) was hit in the spinner and mainplane area. Malan's first burst sent bullets pouring into the Dornier's engines and fuselage, and bits and pieces were blasted off into space. His second attack was made from the beam, curving in to dead astern as his Number Two, Flying Officer Bill Armstrong (P7353), also came in. He opened up from 300 yards with a two-second burst, followed by a six-second burst from 250 yards. He too received attention from the German gunner until he broke away. However, his second burst also hit one of the engines and the fuselage. There was an explosion under the port engine and a great cloud of oil spewed

Wing Commander Sailor Malan DSO & bar DFC & bar.

Pilot Officer H G R Poulton and Flight Lieutenant J C Freeborn DFC & bar, early 1941.

Flying Officer H C Baker and Pilot Officer E A Mould, 1941.

back, covering his windscreen which effectively cut off all forward vision.

Malan made his third and last attack from 50 yards, giving the Dornier the rest of his ammunition. Then he ordered Yellow Section in to attack as Armstrong with his windscreen clearing slightly, formated on his leader's wing. Johnny Freeborn (P7366) made a beam pass from the left as Yellow Two, Pilot Officer Peter Chesters (P7591) came in from the right. Following his second burst, Freeborn saw the Dornier's port engine explode. Chesters noticed that the rear gunner had now ceased all return fire. Chesters and Freeborn continued to pour fire into the Dornier which was now flying only about 200 feet above the sea. Then Chesters, while he was flying and firing from point-blank range, saw one of the German crew climb out onto the port wing, and drop off just as the Dornier hit the sea, smashing into the water in a sheet of silvery-white spray.

The Dornier had certainly taken some punishment in support of Malan's view about its mission. Malan's guns had fired 2,256 rounds, Armstrong's 1,200, Freeborn's 1,360 and Chester's 1,725, even though the latter's No 1 gun had packed up after just 175 rounds. The kill was shared between them. It was also the last victory claimed by Sailor Malan while CO of the squadron.

The squadron flew another sweep on the 11th, between Boulogne and Gris Nez, in company with 66 Squadron. Two Me109s were seen but not engaged, but 66 got into a scrap and lost two aircraft. Johnny Freeborn received a well deserved bar to his DFC in February, just before the squadron moved to Manston.

Of this period, Roger Boulding remembers:

'Other tactics that Sailor was instrumental in introducing (and leading!) during the winter of '40 and the spring of '41 involved much "trailing our coats" over France in an endeavour to persuade the enemy to send up fighters so we would have an opportunity to knock 'em down. The squadron moved to Manston on, I think, 20 February 1941 under the command of Mungo-Park, who was promoted when Sailor became a Wing Commander and the Wing Leader of the Biggin Hill Wing of which we remained a unit.

On one memorable occasion he led us on a beautiful clear day over France with one section flying high enough to make contrails very visible to the Germans – and the other two sections at a much lower altitude – such that we normally would not have contemplated in clear weather. For about twenty minutes we cruised about, expecting a massive anti-aircraft barrage or a sudden fighter attack at any second. In the event the Germans did not react and absolutely nothing happened but they were about the longest twenty minutes I can recall! It was also typical of the man – he was constantly trying to produce ways in which we could persuade the Germans to give us an opportunity to attack them and was never content to just fly over enemy territory with the aim of attacking anybody who appeared. He wanted to provoke them into battle.

I also flew as his Number Two on some of the sweeps across the Channel and found that there was no going back with him until we had done all that could be done – not that he took foolish risks but he was always full of determination – his aim was to get so close behind his target that he couldn't miss. We in 74 Squadron knew him to be a tough and determined and generally a hard man to fly with. His demeanour was quiet and very cool. He seemed to have himself very much under control and I don't remember him ever getting very excited on the ground. Whilst he would often have a few quiet beers in the evening with the rest of us I don't think he was ever a great partygoer and I don't recall him ever going on a binge in the time I knew him. We all, I think, held him in very great esteem and had great confidence in his leadership, even if we were not always too keen on his methods of stirring up trouble.'

On 22 February Pilot Officer E.W.G. Churches and Sergeant Neil Morrison destroyed an Me110, but Morrison was posted as missing two days later. At the end of the month, 66 Squadron left Biggin Hill, being replaced by 609 Auxiliary Squadron, commanded by Squadron Leader Mike Robinson DFC. Mike Robinson was the son of Lord Robinson. He was a brilliant fighter pilot, having already seen action during the Battle of Britain with 601 and 238 Squadrons before taking command of 609 in the autumn of 1940. His very first confirmed victory had been over an Me109 which he forced to land in southern England despite the fact that his guns were empty!

As Boulding mentioned, Sailor Malan had now left the Tigers, becoming the Biggin Hill Wing Leader. He had flown with 74 continually since November 1936, four years and two months. With the new offensive starting, this 'taking the war to the enemy', it had become evident that it was necessary to continually operate above squadron strength. The Fighting Wing concept was born. Yet these concepts needed people to lead them in the air. These leaders must be carefully chosen. They must be born leaders, reasonably good pilots and be able to take charge, in the air, of at least 36 fighter aircraft. A good tactical sense was essential, a pre-requisite far and above the ability to shoot well. If they could, in addition, shoot well, then this was a bonus.

Not unnaturally, however, the first handful of pilots chosen to lead the first wings, had mostly become famous and well known primarily for their ability to shoot and destroy enemy aircraft. The first two men chosen were Douglas Bader, the legendary legless air ace, who was given command of the Tangmere Wing. The other was Malan, who would lead the Biggin Hill boys, comprising his old 74, plus 92 and 609 Squadrons. Soon afterwards Johnny Peel DFC became leader of the Kenley Wing.

Mungo-Park took over the Tigers, whilst Flight Lieutenant Tony Bartley DFC moved across from 92 Squadron to become a flight commander. Johnny Kent DFC left 92 Squadron, his place being taken by Squadron Leader Jamie Rankin, a comparative newcomer to fighter operations, but one who very soon emerged a successful air fighter.

Mungo-Park's promotion was well received by the Tigers. He, like Malan, was a natural leader and born fighter pilot. 'Taffy' Jones recorded in his book *Tiger Squadron* that Mungo-Park once said of Sailor:

> What I like about Sailor is his quiet, firm manner and his cold courage. He is gifted with uncanny eyesight and is a natural fighter pilot. When he calls over the R/T, 'Let 'em have it! Let 'em have it!', there's no messing. The bastards are for it, particularly the one he has in his own reflector sight. Mannock and Malan have made 74.

*

For Sailor Malan it now became a time of preparation and training for the offensive that was about to begin when the weather finally settled down. All his tactical skill was needed and his energy was amazing. He had already proved his ability in this respect but had now to discuss and formulate ideas for operating constantly with three squadrons of Spitfires and in co-operation with small or large numbers of medium bombers, as well as other fighter wings.

A whole new vocabulary began as 1941 progressed and the operations that began to be flown had their own names.

Sweep: General term for fighters flying an offensive mission over enemy controlled territory or sea. Could be flown in conjunction with but not in direct support of a force of bombers.

Rodeo: Fighter sweep over enemy territory without bombers.

Circus: A bomber or fighter-bomber operation heavily escorted by fighters. Designed primarily to draw enemy fighters into the air.

Ramrod: A similar operation to a Circus mission but where the objective was the destruction of a specific target.

Rhubarb: Small scale freelance fighter operation to attack targets of opportunity, usually in bad weather.

Whatever the operation the supporting fighters had to be carefully briefed. There was little sense in filling the sky with fighters who then got in each other's way when enemy aircraft were found. Each operation meant a good deal of careful planning and squadron and wings were given specific tasks and areas of responsibility.

Close Escort: Surrounding and keeping with the bombers.

Escort Cover: Protecting the close escort fighters.

High Cover: Preventing enemy fighters from positioning themselves above the close escort and cover escort wings.

Top Cover: Tied to the bomber route on Circus and Ramrod missions but having a roving commission to sweep the sky in the immediate area of the bomber's course.

Target Support: Independently routed fighters flying directly to and then to cover the target area.

Withdrawal Cover: Fighters supporting the return journey when escorting fighters would be running short of fuel and ammunition.

Fighter Diversion: A wing or wings creating a diversionary sweep to keep hostile aircraft from the main target area on Ramrod operations.

In the meantime, before the full scale operations began, 74 Squadron, as well as other units still flew and fought as the spring weather arrived. On 18 March Pilot Officer Churches got a 109, and Kiwi Spurdle and Sergeant Dales severely damaged lone bomber raiders on the 24th and 25th; then on 7 April, Sergeant J. Rogowski (Polish) got another 109 while Pilot Officer Howard hit a second. Three days later Peter Chesters, who had shared the Dornier 215 with Malan, Freeborn and Armstrong on 5 February, shot down a 109. He was so elated with his success that he decided to fly a victory roll over the station, misjudged it, crashed on the parade ground and was killed instantly. Just over a week later Pilot Officer Churches went down to become a prisoner.

In May the first Spitfire Mark Vs began to arrive. They were an improvement on the Mark II and were to become Fighter Command's standard equipment for a year. As far as hitting the enemy was concerned, the Mark V mounted two 20mm cannon – a vast improvement over the .303 machine-guns. Cannons had been on trial in Fighter Command for many months but early teething troubles delayed their full-scale implementation. Now, with the major problems solved, they were standard equipment for both the Spitfire V and the Hurricane IIC. The new Spitfire still retained four machine-guns, but where they tickled the cannons blasted. Each cannon carried 60 shells, the .303 continued to carry 350 rounds per gun. For a man like Malan these new guns would create a new dimension in combat.

Initially there was some difference of opinion amongst the top pilots concerning the use of cannons. Bader was all for retaining the machine-guns but Malan and others could see the advantages of using the heavier armament. Bader eventually came round but Malan and his followers immediately got cracking with them as soon as everyone had the Mark V.

There were several events in the first part of May 1941 involving the Tigers and 609 Squadron. On the 7th 74 had four pilots on patrol who found ten Me109s flying in loose formation at 20,000 feet off the North Foreland. The four men attacked; Pilot Officer H.G.R. Poulton destroyed one, and Pilot Officer W.J. Sandman, a New Zealander, probably destroyed another, while Roger Boulding knocked pieces off a third.

At 5 p.m. on the 7th, Mike Robinson led A Flight of 609 as escort to a Defiant aircraft spotting for guns from Dover firing on Calais. Failing to contact the Defiant's crew by radio, Robinson appointed one of his Spitfires to broadcast any pre-arranged visible signals the Defiant might give. This was clearly no good, so a Blenheim came out to relieve the Defiant. The Blenheim, however, did not wait for the escort and carried on to Calais with just the lone Spitfire. Both were attacked as Robinson and the others came on the scene, and four 109s attacked Robinson head-on. In reply, Robinson pulled up his nose and fired a long burst at the leading Messerschmitt until he was over on his back. He then started a scrap with two more 109s, hitting one with a five-second burst from behind which produced white smoke and bits flew from it. Robinson then saw a splash, thought it was a hit 109, but then saw it was the Blenheim. The whole thing was a shambles as 609 had no radio contact with the Blenheim crew either.

The very next day six Tigers found more 109s over Dungeness at 23,000 feet, which were again flying in a loose type of formation. 74 were at 30,000 feet and they dived down, scattering the Messerschmitts. Bill Armstrong chased one across the Channel and shot it down into the sea.

May 8 was the birthday of Mike Robinson. He and his squadron had quite a birthday party that afternoon. Shortly after 5 p.m., 609 were scrambled to patrol Maidstone at 15,000 feet. Robinson was then advised by radio that there was a dinghy in the sea halfway across the Channel and given a course vector of 100 degrees. Flying out he saw a rescue boat on fire off Dungeness with two Me109s circling it. The 109s flew off before they could be engaged, but they then saw two other 109s near the burning vessel. Leaving Yellow Section as top cover at 10,000 feet, Robinson led Red and Blue Sections down. He attacked one of the Messerschmitts and shot it into the sea with a five-second burst. Chasing the second 109 towards France he began firing at 800 yards down to 50. White smoke blew back and the 109 landed in the sea just short of the beach. At once nine more 109s came down on him but Robinson was no novice and successfully dog-fought his way out of trouble.

Meanwhile Flight Lieutenant Johnny Curchin (P8264) led the others in a scrap with 109s, and the final claims totalled, with Robinson's two, six destroyed plus two probables. Curchin got one and shared another with Sergeant Hughes-Rees (P8271), Sergeant Tommy Rigler destroyed two (P8422) while Sergeants Mercer and Palmer (P7734 and P7625) got the probables.

Two days later Roger Boulding shot down a night raider:

'In May 1941 we moved to Gravesend, still as part of the Biggin Hill Wing and it was whilst there that we were involved in reinforcing the night defences of London during the three days of the full moon period, considered to be the most likely time for a major night attack.

'The idea was that the AA barrage would be limited over the central area of London to a height of about 12,000 feet. Outside this the guns were permitted to fire up to their maximum and beyond the barrage were out night fighters operating under ground control. Day fighters were to patrol over the central area at altitudes above say 15,000 feet, individually and at 500 foot intervals. We were to gain access to this central area via a gap in the barrage marked by two coloured searchlight beams (blue I think) fixed in the vertical position. This device was almost entirely non-effective and I think most just flew through the barrage to our appointed patrol areas.

'Night flying in a Spitfire was not a very satisfactory proposition for a number of reasons but the main problem was the glare from the exhausts on either side of that long nose. This effectively reduced one's night vision to a rather nominal level.

'On 10th May the sky was clear of cloud and visibility very good. The Germans caused a major conflagration and from the air London seemed to be at the base of a pyramid of flames – a truly horrifying sight, visible to us just as soon as we were clear of the ground. I think we all did two sorties that night (from West Malling airfield) and we could see that the defences were having some success because we saw quite a few aircraft go down in flames, but we had little chance ourselves and I believe not one of the day fighter pilots that night, except me, even saw an enemy aircraft.

'I was on patrol over the central area and a little before my scheduled time to return to base, experienced a problem with my constant speed unit which automatically controls the propeller. This wasn't too serious but made it advisable to cut short the trip. I had established radio contact with West Malling and just set course for home when I suddenly saw by the light of the full moon that I was just behind and below a large twin-engined aircraft – unmistakably an enemy who was returning from the target area. It was only necessary to raise the nose slightly and press the gun firing button. The stream of bullets from all my eight machine-guns converged on the underside of the Heinkel 111 and an enormous cloud of sparks dazzled me briefly. A rapid search revealed that the He111 was diving quite steeply towards the blackness of the ground (from around perhaps, 15,000 feet) but he was illuminated by the brilliant moon. I dived after him but every time I got near a position to open fire again I was silhouetted against the lighter sky. I never did get into a good firing position again, and had my work cut out to keep track of him through the glare of my exhausts. When he finally disappeared from view I circled the area and obtained a radio fix, then returned to base where we found the guard over my oil-cooling air inlet had been hit by debris and, later, that a He111 had been located near the radio fix position (at Kennington, near Ashford, Kent).

'A few days later some of us drove down to the site and found that the German pilot had managed to make a wheels-up landing of sorts in a small field. There was much evidence of bullet strikes under the fuselage etc; but otherwise the aircraft was still fairly complete. We discovered the crew had been taken to Chartham Hospital near Canterbury but were unable to see them because of their injuries. I still have their signatures on a piece of paper that one of the medical orderlies got for me.'

As the May weather improved, the long awaited chance to mount regular offensives operations over France finally came. Malan led 74, 92 and 609 Squadrons off on a fighter sweep shortly before mid-day on the 16th. Several Me109s were seen and Flight Lieutenant Alan Wright DFC and Sergeant Don Kingaby DFM , both of 92 Squadron, scored probables over the Channel. They landed back at 2 p.m., and another patrol was flown in the late afternoon; Kingaby destroyed a 109, and a second was blasted by a number of pilots of 92 Squadron.

The following day, the 17th, Sailor led a high patrol of six Spitfire Vs from Dungeness to the North Foreland at 32,000 feet. He reduced height to 23,000 feet over Dover on receiving information from base, and at 12.50 p.m., he spotted six Me109Fs flying at 20,000 feet. They were in a dive from the north, flying from the direction of Canterbury and had clearly been up to no good.

The six Spitfires half-rolled after them, chasing them out to sea but they could not get closer than 800 yards. Then four of the Germans pulled up into a climb while the other two dived towards France. Malan pulled over to their starboard flank but found himself out-positioned by the two Messerschmitts and was forced to withdraw. Returning towards Dover, he then found and surprised four Messerschmitts flying in pairs 2,000 feet apart, one pair at 18,000 feet, the other two two thousand feet higher.

Malan attacked the lower pair on a steep turn to the right, giving them both several bursts. His last deflection burst hit one of the Messerschmitts around the cockpit and engine. Its hood or part of the engine cowling, whipped back, and then most of the 109 became obscured by black smoke. However, the top pair had seen the attack and were coming down, firing, forcing Malan to break away. The last he saw of the 109, it was on its side, throwing out heavy smoke, with other 109s circling it protectively.

As his engine was vibrating very badly, Malan called it quits, and landed at Hawkinge, returning to Biggin Hill by Magister. He was allowed a probable – his first claim as a Wing Leader.

Four days later, 21 May, Circus Numbers 10 and 11 were organised but weather affected the plans and in the event only Circus 10 was flown in the late afternoon. Eleven Blenheims from 21 Squadron and six of 110 Squadron were briefed to bomb the oil refinery near Béthune and they were supported by almost every fighter squadron within 11 Fighter Group, plus other units from 10 and 12 Groups. The whole operation was kept a closely guarded secret from all the pilots until shortly before take-off. For Biggin Hill and its Wing Leader, Sailor Malan, take-off came at 5.10 p.m., Malan leading 609 Squadron, acting as cover for the Kenley Wing whose squadrons were providing the Blenheim's close escort.

The target was reached and bombed successfully. Apart from hitting one bomber of 110 Squadron, anti-aircraft fire did little further damage to the attacking force. It was on the way out that the Luftwaffe appeared. Meanwhile, Malan and 609 patrolled ten miles east of Deal on the line Gravelines–North Foreland in three sections, in line astern, between 12,000 and 14,000 feet. As the Circus aircraft came out over Gravelines at 6 o'clock, Malan was near North Foreland. He saw several small formations of Me109s flying in fours and pairs, mostly at the same height as his Spitfires. As he turned, two Me109s appeared on his left, below, and two more to his right, but turning to the right.

Malan found his radio had failed, so signalling to his Number Two, Pilot Officer Peter MacKenzie (P8369), he dived to attack one of the left-hand Messerschmitts; his wingman went after one to their right. However, MacKenzie blacked out in the turned as he pressed his gun button, and spun away. Three of 609's Belgian pilots followed him down, leaving the Wing Leader alone. Malan continued after one Messerschmitt, gave it three bursts with slight deflection from 300 yards, following this with one three-second burst from dead astern in a dive at 200 yards range, which caused smoke to stream back from the German fighter.

Then three 109s on his right turned in towards him, and Malan turned towards them in order to evade their attack. Alone again for a moment, he saw three Spitfires ahead and flew towards them, but then four more 109s made a pass. Malan evaded these by spiralling down, then pulling up into the sun, and then diving. The four 109Fs stayed with him, chasing him in a straight dive and although he was in a Spitfire V, they were clearly catching up, so at 9,000 feet Malan had to turn and then spiral down and away. He landed back at Biggin Hill at 6.30 p.m.

The others in 609 also got into several scraps with the 109s during a running dog-fight. Sergeant David Hughes-Rees damaged one while Pilot Officer Vicki Ortmans, a Belgian, forced another down which was then attacked by four Spitfires. One of these, Flying Officer John Bisdee, fired a burst and saw it go down in a slow spiral with little sign of it recovering. Then Bisdee saw tracer passing him and immediately saw pieces of aircraft in the air and a parachute descending. This was his Number Two, Pilot Officer Comte Rudolphe de Grunne, who was surprised and hit by a 109. Yellow Three and Four, Sergeants Palmer and Boyd, circled de Grunne; Palmer stayed with him until he hit the water. Although a rescue boat set out from Ramsgate, he was not found, and Palmer had to leave the area when his fuel ran low.

De Grunne was one of 609's beloved Belgians and had only recently returned to operational flying after having been shot down and wounded while flying with 32 Squadron in the Battle of Britain. He had previously flown with the Nationalists in the Spanish Civil War, flying Fiat CR32s and later Messerschmitt 109s, destroying 14 Republican aircraft. Before joining 609 he had been to Lisbon on a secret mission, picking up gossip about the possible invasion of Spain and North Africa. He came to 609 sporting a superb suntan. The last thing he did before taking-off on the 21st was to rush back to the Squadron Intelligence Officer, as he had left his lucky miniature horse-shoe in his wallet.

CHAPTER SEVEN

Wing Leader – Biggin Hill

Wing Commander Sailor Malan and Squadron Leader Jamie Rankin flew a Rhubarb operation in the late afternoon on 25 May, Malan in R6882, Rankin in R7346. The weather was bad with rain and strong wind. They found nothing. Flying Officer H.C. Baker of 74 Squadron had better fortune the next day, bagging an Me109 on another Rhubarb mission.

Early in June, veteran Tiger Sergeant William Skinner DFM left the squadron, but returned four days later with a commission as a pilot officer.

Then on 7 June, Malan (R7192) led a fighter sweep over the French coast. It had started out a dull day but with the weathermen's promise of improvement, the pilots came to readiness at 11.30 a.m., taking to the air fifteen minutes later. They flew for 1¼ hours but the Germans stayed on the ground, so the Spitfires landed empty-handed. Five days later, the 12th, Malan led 74 and 92 as escort to three Blenheims, meeting the bombers over Canterbury.

Take-off from Biggin was at 2.19 p.m., then with the Blenheims below them the Spitfires headed out towards Dunkirk. Malan and 74 led them in to bomb shipping while 92 patrolled over the coast. Eight Me109s came up, and six of them attacked 92. Jamie Rankin shot down one into the sea. Four days later it was the Germans' turn to hit back.

In a sweep, a scrap developed with a bunch of 109s and although the wing claimed four Messerschmitts plus two probables, Mungo-Park limped back streaming glycol after getting two 109s, to crash-land near Hawkinge, while Pilot Officer Parkes and Roger Boulding were lost. Boulding (W3521) relates:

> As we were his 'own' squadron he [Malan] normally led his wing with 74 Squadron. Sailor was leading us on the trip

.***(Left)*** Roger Boulding and 'Sam' in front of his Spitfire P8394 ZP-S, 1941. ***(Right)*** Flying Officer R J E Boulding who often flew as Malan's number two until he became a prisoner of war in June 1941.

(Left) Pilot Officer W J Sandman, another wingman of Malan's until he too was taken prisoner in June 1941. Spitfire P8261 ZP-N. ***(Right)*** Pilot Officer W J Sandman aged 20, from Auckland, New Zealand.

(Left) Sergeant W J Lockhart, another Number Two to Malan in 1941. He was reported missing in action 6 July 1941. ***(Right)*** Neville Duke learned his trade as a fighter pilot with 92 Squadron in 1941, often as Number two to Sailor.

during which I was shot down. It was a clear day and we were well across the French coast. I was leading one section of four and spotted a formation of Me109s (with the famous yellow painted noses) climbing towards us. I radioed the sighting to Sailor who led us in a diving turn straight onto them. I followed one down in a near vertical dive but had to break off without apparently causing him major damage. At that time we had strict orders not to pursue down to low level over the other side – and Sailor was radioing us to reform. I tagged along some little way behind him, both of us using the familiar tactic of flying towards the sun in a weaving pattern so as to present a difficult target.

I looked behind and spotted another Spitfire following me in the same fashion. Shortly afterwards Sailor began to call for someone to 'look-out behind!' and urgently to take evasive action. I looked behind, saw what I thought was the same aircraft guarding my rear and began to hunt round for the one in trouble. I had just spotted a Spitfire rocking its wings violently (probably Sailor) when my aircraft was hit from behind (the armour plate behind my seat took it and saved me). The aileron controls went and the stick just flopped from side to side without effect. My aircraft went into a spiral dive, starting from about 25,000 feet, and I had to get out fast. I pulled the canopy release without too much trouble, undid my seat straps but could not get out because of the spinning so had to get my knee up and jerk the stick forward, which effectively catapulted me out. I pulled the ripcord and parachuted down from, at a guess, somewhere above 10,000 feet. The Germans had ample time to reach me when I landed and before I could stand up there were plenty of them threatening me with an assortment of weapons.

*

The following afternoon, 17 June, came Circus 13. Perhaps at this stage we should look in some detail at exactly how a Circus operation was organised and what was involved.

The operational orders for Circus 13 called for 18 Bristol

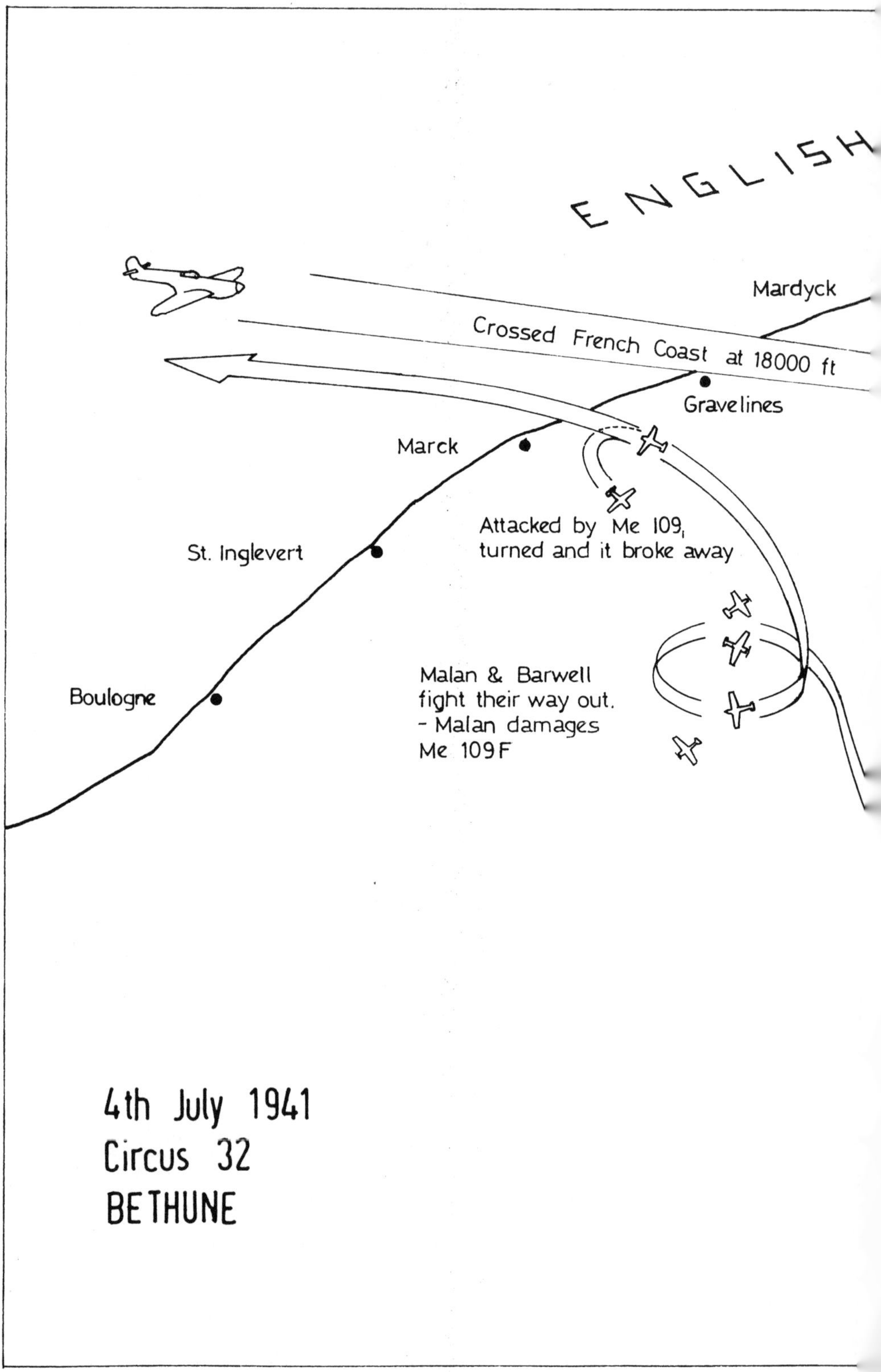
ENGLISH
Mardyck
Crossed French Coast at 18000 ft
Gravelines
Marck
Attacked by Me 109,
turned and it broke away
St. Inglevert
Boulogne
Malan & Barwell
fight their way out.
- Malan damages
Me 109F
4th July 1941
Circus 32
BETHUNE

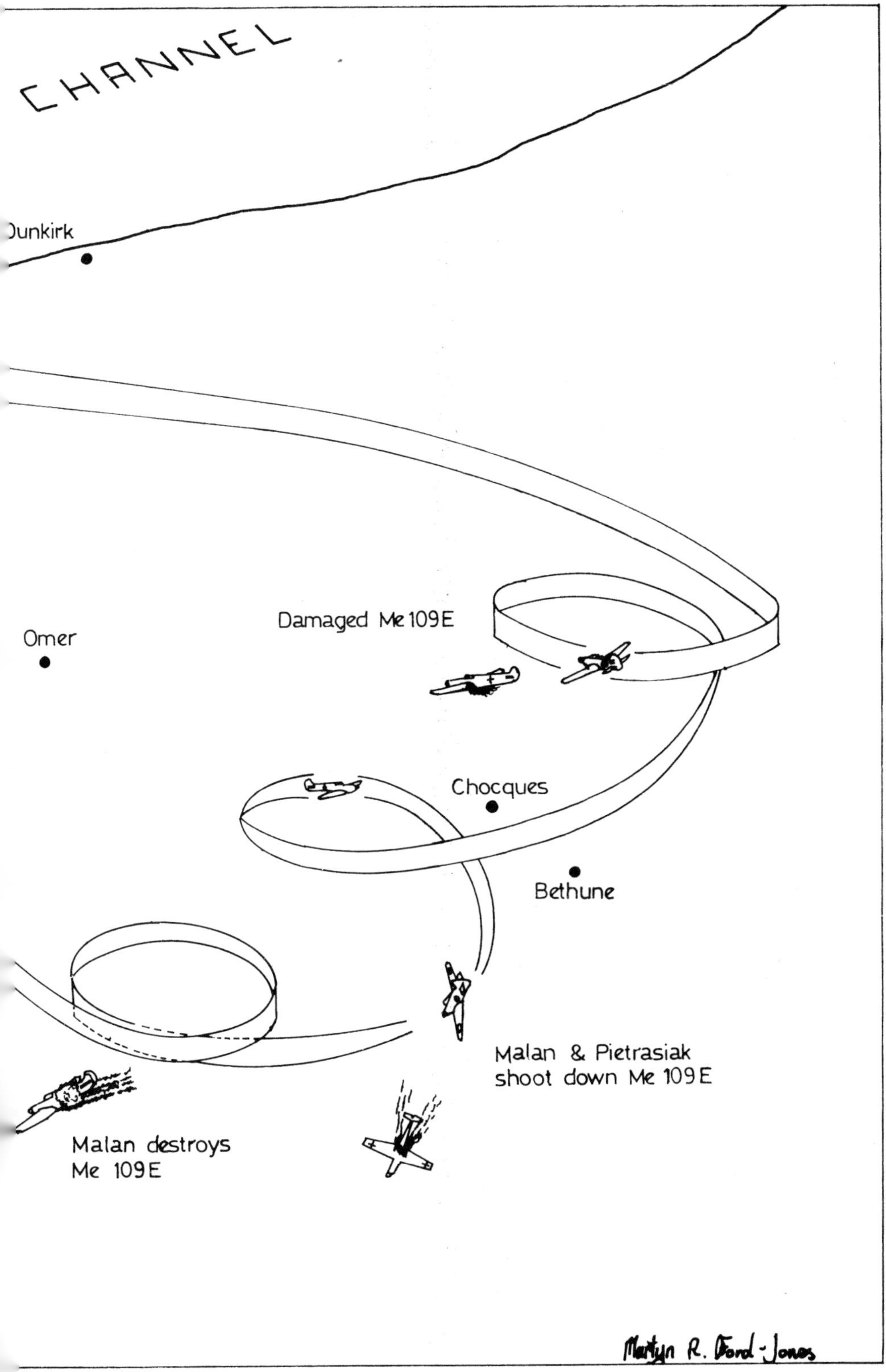
CHANNEL
Dunkirk
Omer
Damaged Me 109E
Chocques
Bethune
Malan & Pietrasiak
shoot down Me 109E
Malan destroys
Me 109E
Martyn R. Ford-Jones

Blenheims from 2 Group Bomber Command, to bomb the chemical works at Chocques, near Béthune. Seven Blenheims came from 18 Squadron, six from 110 Squadron and five more from 107 Squadron. The Close Escort Wing was provided by 56, 242 and 306 Squadrons, the North Weald Wing, while Malan's three squadrons from Biggin Hill were assigned High Cover Wing. Five fighter squadrons, 303, 54, 603, 611 and 91, who were mainly from the Hornchurch Wing, were the Main Fighter Force of the Forward Echelon, while the Rear Escort squadrons were 1, 258 and 312, the Kenley Wing, plus the Tangmere Wing of 145, 610 and 616 Squadrons, who flew an offensive fighter sweep. Support Wings came from 10 Group, 308 and 501, and 12 Group, 19, 65 and 266 Squadrons.

The visibility over Northern France was excellent above a ground haze which reached an estimated 2,000 feet. There was no cloud, just a vast expanse of blue sky.

Chocques meant quite a deep penetration of enemy air space at this stage of the air war, and with 18 Blenheims and upwards of 250 fighters somewhere in the air, Circus Number 13 was easily detected by German radar. Enemy fighter opposition proved quicker off the mark in consequence. For once the Germans were really reacting to the Royal Air Force's incursions.

The Close Escort Wing was engaged almost continually after they crossed the French coast, right up to the target. Yet they prevented the Messerschmitts getting in amongst the Blenheims and without loss, claimed three 109s destroyed. As the bombers completed their attack, 56 Squadron, flying 2,000 feet above and behind the bombers, turned to engage three Me109s that had been shadowing them, forcing them away. Then 56 were continually attacked by about 20 Me109s until they had reached about mid-Channel on the way out, losing four pilots, although one was heard over the radio saying that he was about to crash-land in France.

Number 242 Squadron's B Flight, part of the second escort squadron, was attacked by three Messerschmitts when about 15 miles from the target, but they broke away when A Flight turned after them. 242 were then continually engaged by 109s and breaking into pairs, they fought a rearguard action right out into mid-Channel where the Messerschmitts finally left them. They had been engaged by an estimated 40-50 109s and claimed one destroyed, three probably destroyed and nine damaged, but they lost three of their Hurricanes.

Meanwhile, the Biggin Hill Wing flew above, 74 and 609 on both flanks of the bombers, 92 flying as High Cover above. They had flown out over Southend at 15,000 feet and when they crossed the French coast they were stepped up from 24,000 feet, Malan keeping them at this height as ordered until just before reaching the target area. Malan then descended to 19,000 feet and, as the Blenheims turned for home, he saw seven Me109Fs slowly overtaking the Close Escort Hurricanes who were flying at 12,000 feet.

Malan gave a warning over the R/T, and then took eight of 74 Squadron down to attack. During the dive down, several single 109s were encountered and were engaged by Malan's sub-sections. Then the original seven Messerschmitts saw him and half-rolled away. Malan followed to about 10,000 feet but then decided to regain height in order to support the middle layers at 15,000 feet. Seeing a 109 chasing a Spitfire, Malan tried to get onto its tail but then he became entangled with two other 109s. One half-rolled away but he continued to turn with the second in a left-hand circle. He snapped off several short bursts, but they all found empty air just behind the German's tailplane. Then the 109 straightened up and he hit it with a burst of cannon which hit the Messerschmitt around the engine, port wing and cockpit. Then its port wing sheered off and the luckless 109 spun away at a terrific rate with its engine full on, pouring out black smoke. As Malan watched the spinning fighter he saw it erupt into a mass of flames.

He continued to climb towards the coast and when he reached 20,000 feet met a 109F and circled briefly with it, before it straightened out and began to draw away from his Spitfire. Malan tried to follow and even engaged his emergency boost control but it continued to fly away. As he tried desperately to catch up, he saw the 109 approach a lone Spitfire, and yelled out a warning. The Spitfire merely went into a gentle turning movement as the 109 closed in. When Malan was still 2,000 yards away and unable to do more than continue to shout warnings to the Spitfire's pilot, the 109 opened fire. The Spitfire flicked into a spin and went down, he saw a parachute appear suddenly below. Still trying to close with the 109, it, in turn, did a half-roll and dived to safety.

Frustrated, Malan turned for home, crossed out of France just south of Boulogne and picked up the rest of the wing over the English coast. He found upon landing that his starboard cannon had packed up after just five rounds, the port cannon having fired all its 60 rounds.

Tiger Squadron lost two pilots, while 92 Squadron fought 20 Messerschmitts destroying two for no loss. 609 were first engaged when nearing the French coast on the way out, two or three pairs coming in from the north. Flight Lieutenant Bisdee's cannons jammed but using just his machine-guns, sent one 109 down 'like a torch'. Flying Officer Keith Ogilvie caused another to blow-up and hit the sea. Pilot Officer Francois de Spirlet, a Belgian pilot, sent another down pouring out smoke and later Mike Robinson saw a patch of frothing water below the action.

The main fighter force of 54, 603 and 611 Squadrons, had left the English coast at Dungeness and flown on a zig-zag course out into the Channel at between 18,000 feet and 20,000 feet. They eventually broke into sections, seeing the bombers and close escort coming out of France. Patrolling and sweeping to and fro off Cap Gris Nez, they saw no sign of the opposition but then 54 met 109s and claimed two destroyed and two probables for no loss.

Number 91 Squadron patrolled at 19,000 feet ten miles off the French coast between Calais and Boulogne, and saw two Me109s, but they dived away when 91 appeared. They saw two aircraft crash into the sea off Gris Nez and later escorted a Spitfire which was losing coolant, but it could not make it and went down into the sea, its pilot stepping out of the cockpit after a successful belly landing. Not an easy thing to do in a Spitfire. The Poles of 303 Squadron failed to make rendezvous with the Circus, so flew to Gris Nez independently getting into a scrap with Messerschmitts; they claimed one, plus a probable. One of the Polish machines was hit in the radiator which forced the pilot to crash-land near Manston.

The rear echelon of the Main Fighter Force, 1, 258 and 312 Squadrons, patrolled as far as mid-Channel between Dover and

Dungeness; 258 and 312 Squadrons only met intense flak but 1 Squadron, flying in sections, saw several 109s dive straight through the bombers and on towards France. One pilot chased after one of the Messerschmitts and destroyed it.

The offensive sweep carried out by 145, 610 and 616 Squadrons went into France at Le Touquet, flew towards Guines and then flew out between Calais and Gris Nez. Their brief was to take on any fighters they might find between the bombers and the Main Fighter Force. Arriving a few minutes later, the Wing Leader cut the corner and flew directly to Boulogne and Calais. Enemy fighters were met well out to sea and one pilot in 610 Squadron claimed one destroyed. Both support wings from 10 and 12 Groups patrolled over the Channel but made no contact with hostile aircraft.

So ended Circus 13. The Royal Air Force claimed 15 Messerschmitt 109s destroyed, seven probably destroyed, plus 11 damaged. The RAF lost 11 pilots. Malan had blasted a wing off of a Messerschmitt, the first kill of an incredible rate of scoring that he was about to begin, while his wing destroyed a further five but lost two pilots.

As can be seen, a Circus operation could be fairly involved and to describe it as a few medium bombers escorted by several squadrons of fighters is something of an over-simplification. For Wing Commander Malan and his Biggin Hill Wing, these types of operations were soon to become an almost daily occurrence over the next few weeks.

*

The operational diary of 609 Squadron recorded that for them, 21 June 1941 was the biggest day since the Battle of Britain. It was also the longest day of the year, the weather was the hottest so far recorded that summer and it was also the day the Germans invaded Russia.

Two Group of Bomber Command and 11 Group's fighters put on Circus Numbers 16 and 17. Circus 16 comprised six aircraft from 21 Squadron which attacked the airfields at St Omer/ Longuenesse and Fort Rouge, but it was Circus 17 which caused

Biggin Hill 1941, l. to r.: John Bisdee DFC, Paul Richey DFC of 609 Squadron, Group Captain P R 'Dickie' Barwell DFC, who flew several missions as Malan's Number Two while Station Commander of Biggin, and 609's CO Mike Robinson DSO DFC.

609 Squadron, Biggin Hill, 1941. l. to r.: Standing: Bob Boyd, Bauduoin de Hemptinne, Peter MacKenzie, Paul Richey, John Bisdee, Jean Offenberg and Jimmy Baraldi. Seated, Vicki Ortmans, Tommy Rigler, Keith Ogilvie, and Bob Wilmet.

609 to make mention of this day in their diary. Six Blenheims from 110 Squadron were detailed to bomb Desvres airfield, and Malan's wing acted as target support wing. Take-off came at 3.55 p.m.

The Biggin Hill Wing followed the same route as the bombers; Malan led them over the English coast above Deal in Kent, then flew on a course to Gravelines, crossing the French coast on a dive at 22,000 feet, heading straight for Desvres. The weather was so hot that many pilots flew without their tunics or flying jackets. The bombers reached the target at 4.31 p.m.; Malan's wing, flying in fours at 15,000 feet, stepped up to 18,000 feet, saw them coming in.

Then Malan spotted a Messerschmitt 109F flying towards Boulogne at 15,000 feet, attacked and closed in to 50 yards before firing two short burst of cannon and machine gun. The 109 burst into flames as oil spewed back to coat Malan's Spitfire. His Number Two, Pilot Officer W.J. Sandman, a New Zealander with 74 Squadron, saw it go down on fire just inland of Boulogne.

Already Malan's keen eyes had picked out six more 109Fs at 12,000 feet off Le Touquet. Four of these began to dive away towards the land, leaving the other two flying peacefully along. Malan winged over and dived towards them, selecting the right-hand Messerschmitt as he came in from the right. He gave it a short deflection burst but missed. The 109 straightened up, allowing Malan to fire from dead astern at a range of 50 yards. The German single-seater jerked violently forwards to dive vertically into the sea at the mouth of the river at Le Touquet. Malan then returned, meeting up with the bombers and their close escort over Dungeness.

Meanwhile, Yellow Section of 609 Squadron, when returning to the French coast between Hardelot and Le Touquet, saw four Me109Fs flying on a parallel course to themselves. As they converged, Yellow One, Keith Ogilvie, turned head-on and fired, thus starting a general dog-fight. Ogilvie latched onto one Messerschmitt and shot it down near Le Touquet; the pilot baled out. Sergeant Boyd fired at another 109 which overshot its dive and spiralled down from 3,000 feet to crash in a field.

Pilot Officer Ortmans, after shaking off one German fighter, went after another which was attacking a Spitfire. Following two beam attacks the 109 dived away streaming black smoke, but with his own engine starting to vibrate, he had to break off the action. Ogilvie was later attacked by four 109s and after some difficulty shook them off although his Spitfire was twice hit in the wing. Thinking he could see Dungeness below he then realised it was Boulogne and had to fight more 109s before finally getting away. Jean Offenberg, another Belgian, chased a 109 to Le Touquet, attacking it from head-on, but low fuel forced him to turn for home without pressing the attack further.

One of the German casualties of 21 June was the *Kommodore* of Jagdgeschwader 26, Oberstleutnant Adolf Galland. During the day he had twice been in action, claiming the destruction of two Blenheims around mid-day before his Messerschmitt 109F-2 was hit by fire from a Spitfire, causing him to belly-land on Calais-Marck airfield. Flying in another 109 against Circus 17 in the afternoon, he shot down a Spitfire which hit the ground a few miles south-east of Boulogne – Galland's 70th victory of the war. Then for a second time he was hit by the fire of a Spitfire and again his fighter was badly damaged and he himself was wounded in the head and right arm. Taking to his parachute and being burnt slightly as he did so, he also dislocated his ankle on landing heavily. It will be remembered that Ogilvie saw the pilot of his victim bale out, but whether this was Galland is open to debate.

The RAF that June day claimed 15 German fighters shot down, three probable kills and two damaged. They lost only a handful of their aircraft although the Germans claimed 14 RAF machines.

Strangely enough, Adolf Galland's ancestry, like Malan's, can be traced back to the French Huguenots. Whereas Malan's forefathers went to South Africa, Galland's chose Westphalia!

During the following afternoon, near Dunkirk, Sailor Malan got another 109F, but it was his wingman who claimed the major honours. This was again Pilot Officer Sandman, aged 20, from Auckland, New Zealand, a member of the Tiger Squadron. They were flying on a cover patrol behind Dunkirk and had climbed to 25,000 feet, then upon reaching their assigned patrol area, came down to 20,000 feet. Two Me109s were seen five thousand feet below flying towards the French coast. 74, led by Malan, dived to attack, following the 109s down to a height of 7,000 feet. Sandman got in a burst from 250 yards but then had to break away owing to the nearness of other Spitfires. Malan however, saw this 109 spin away, burst into flames and hit the ground. Regaining height through some cloud they encountered flak just north of Gravelines, although most of it exploded well behind them.

'Sandy' Sandman was weaving gently behind Malan and another Spitfire about four miles from the French coast when six Me109s attacked him from behind. He pulled round in a very steep turn, and the 109's fire passed behind him. Sandman blasted one Messerschmitt from 50 yards with cannon and machine-gun, seeing large flashes on its fuselage while bits and pieces flew off. The 109 lurched over onto its back and went down, straight into the sea below.

June was becoming hectic as the pace continued. Malan was in action for the third successive day on the 23rd during Circus 20. The weather was again bright and clear. 92 Squadron had flown a convoy patrol early in the morning and the wing then flew an escort mission to Circus 19, in the early afternoon. Jamie Rankin destroyed two Me109s and Don Kingaby and another pilot destroyed two more.

Ninety-two Squadron was the middle squadron in the early evening mission when Circus 20 was mounted. Six Blenheims from 107 Squadron were briefed to bomb the airfield at Mardyck, right on the enemy coast.

Malan led the wing off at 7.47 p.m., being the operation's support wing. By the time the French coast was reached at 8.20, they were flying at heights between 25,000 and 27,000 feet. Malan led them inland towards St Omer, a feint to draw off enemy fighters, and saw red anti-aircraft puffs which he took to be sky markers. Still heading towards St Omer he picked out a formation of 15 to 20 aircraft below at around 20,000 feet, in a large vic formation and in pairs, coming from Boulogne towards Hardelot, obviously to investigate the red smoke puffs.

Switching his microphone on, Malan called to 74 and 92 Squadrons to follow him as he turned towards the aircraft which he could now recognise as Me109Fs. 74 failed to hear his message, and only Rankin's 92 Squadron followed. Malan pushed the stick forward and eased the throttle forward, powering down to select a 109 flying on the right of the formation, closing in to his usual range of 50 yards before caressing the gun button, sending a short burst of cannon and machine-gun into the German fighter. In a split second and right in front of him the Messerschmitt simply blew-up in a ball of flame, and Malan flew through the resulting shower of whirling debris. This debris hit and damaged his port wing and starboard wing-root and on returning to Biggin he discovered that one piece of jagged metal was still wrapped around the pitot head.

Narrowly missing one large piece of the 109's fuselage that seemed to hurtle narrowly by him, Malan saw another 109 climbing up-sun on his right. He hauled round after it,sat on its tail at 50 yards, thumbed the gun button to pump a two-second burst of cannon and machine-gun fire into it. Mortally hit by the South African's deadly fire, the 109's wing rose as the aeroplane went into a half-roll before nosing over into a headlong dive, black smoke staining the blue sky behind it. A quick check for danger behind him and he was following the German down, firing into the Messerschmitt until his guns fell silent at 8,000 feet, with his airspeed then registering 450 mph. Pulling out but still watching the 109 he saw it begin to disintegrate at 5,000 feet, its wings breaking off, while the fuselage crashed to the ground some ten miles southwest of Boulogne.

Pilots of 92 Squadron saw both of Malan's victories, Pilot Officer Dougall seeing the tail of the second 109 also rip away, Malan not actually seeing that as his own engine cowling obscured the sight. 92 also claimed a 109 and two more as probably destroyed; one of these fell to Sergeant A. Le Cheminant, flying in Malan's section.

*

Although the squadrons were up early the next day – 92 was at dawn readiness – it was not until evening that Circus 21 was launched. Malan led his wing away from Biggin Hill at almost exactly the same time as the previous day. The three squadrons made rendezvous over West Malling at 10,000 feet, with 92 Squadron below, the Tigers 2,000 feet above, and 609 above them as top cover. The wing was assigned as cover squadrons and climbed through cloud to the North Foreland. Here they met up with Bader's Tangmere wing at 8.15 p.m., then proceeded onto course 136 degrees magnetic for the target. Somewhere below seven Blenheims of 18 Squadron, six of 107 plus a further five of 139 Squadron, headed out towards the power station at Comines. Visibility over France was poor, with traces of haze up to 18,000 feet.

There was no opposition on the way to the target which the wing itself reached at 8.35, commencing to fly a large circle around it. 92 Squadron was just below Malan at 20,000 feet. Leaving Comines, Malan followed what he could see of the Tangmere boys back towards the French coast which they made at Dunkirk. At no time in the operation had they seen either the Blenheims or the close escort fighters.

Sailor turned his fighters in a circle between Dunkirk and Gravelines for some time and encountered several formations of Me109Fs which were above them at about 25,000 feet. The wing split into sections of four; Malan pulled up into a series of climbing turns against one gaggle of six Messerschmitts over Dunkirk, but the 109s were not playing and kept above the Spitfires, making no attempt to attack the RAF machines.

Getting nowhere with these Germans, Malan broke away when he saw a further six 109s, five miles to the north-west, but again the German pilots refused to take up the challenge. Then below, through the haze, he saw the main Circus formation crossing out over the coast at Gravelines. He radioed to his men that he was joining them.

Positioning himself about 4,000 yards behind the top rearmost close escort squadron at about 19,000 feet, he saw three Me109Fs sudeenly dive down from the left flank; one of them was heading for a section of Spitfires, its guns already flashing. The 109s then turned to the right, with Malan following in a climbing right-hand turn to fire a two-second burst from 50 yards at the leading Messerschmitt. It was the same 109 which had fired at the

Spitfires. Malan's fire hit it underneath the fuselage, and two large pieces of Messerschmitt drifted lazily away. By now Malan was almost on his back and in a turn of a spin, only had a brief second to see the 109 begin to emit heavy black oily smoke and flicking into a spin towards Gravelines. Straightening out in a dive, another 109 came in to fire at his Spitfire but it then flashed across his nose and was away.

Back at Biggin, 92 Squadron confirmed the Wing Leader's victory and as Sergeant W.J. Payne (W3308) had shot down two 109s they could celebrate. 92 Squadron's total victories had reached 150!

*

There was no let-up in the June offensive, two Circus operations, Numbers 22 and 23, being flown on the 25th. Circus 22 was scheduled for the morning, a readiness state existing at 9.10 a.m. Malan's wing crossed the English coast with all three squadrons flying as top cover at 30,000 feet. This height was maintained to the target area, then as they turned over the target, the marshalling yards at Hazebrouck, they saw the target support wing approaching from the flank. Letting down slightly to 28,000 feet, the Biggin Hill wing split into sections, flying a large 'S' shaped circuit over the area between the French coast and the target.

Sailor's ever watchful eyes saw the enemy first; two Me109Fs 3,000 feet below. Leading his section down and seeing one of his pilots attack from close range, Malan closed with the other but in checking the surrounding sky, saw two other Messerschmitts closing in. Breaking away, he turned to attack the two newcomers. Coming round in a screaming turn behind the two enemy fighters he allowed for slight deflection and opened up at 250-300 yards, firing all his cannon ammunition. However, his reflector sight was not functioning properly and his aim was way off.

With his heavy armament gone, Malan continued to stay with his adversary but the 109 pilot kept putting his nose down, Malan's engine cutting out momentarily as he tried to follow it. Eventually his gun-sight came on target again and, allowing this time for error, Malan closed to 50 yards, let go with his machine-guns, and started the 109 smoking heavily.

Then the second 109, which had been flying in line abreast, broke away at 5,000 feet and began to turn. Malan broke away from the crippled Messerschmitt to meet it head-on. As they flashed passed each other, Malan hauled his machine round in a tight turn and went for the 109 again, but then the 109 was over, down and away. Malan saw a large explosion and fire to his left about one mile away which appeared to be his first victim hitting the ground.

Ninety-two Squadron were again in evidence. Flight Lieutenant Alan Wright (W3265) destroyed a 109 while Pilot Officer Neville Duke (R6904) got a second. Neville Duke, later to become a successful fighter pilot with the Desert Air Force in North Africa and Italy, and later an even more famous and successful test pilot, often flew as Number Two to Sailor Malan in the summer of 1941. He remembers Malan well:

'When I was lucky enough to be posted to 92 Squadron of the Biggin Hill Wing in April 1941 the name and reputation of the Wing Leader, Sailor Malan, was already a byword to the pupils in the Operational Training Unit – our instructors having either served with or known him during the Battle of Britain.

With a grand total of 145 flying hours and fully 26 hours on Spitfires, I was a very shiny new and green pilot officer aged 19 and one of the first two wartime trained recruits to 92 Squadron. During my time the Biggin Hill wing was led by Sailor Malan until early September 1941 when his place was taken by our CO, Jamie Rankin. It was the practice of the Wing Leader to fly with each of the squadrons (92, 74 and 609) on the fighter operations taking place over Northern France during that period – fighter sweeps, close and top cover to bomber operations, withdrawal cover, delousing sweeps and rear support.

'By July I suppose I was considered a veteran of 19 and respectable to fly Number Two to the great man on many occasions when he led 92 Squadron. It was a daunting prospect and at all costs one was to stay with the Wingco no matter what – not only to ensure his cover but the honour of 92 was at stake! Better not to come back at all rather than lose the Wingco.

'This was all right whilst he was leading the wing peacefully about our business – he was a master at the art of leading some 36 aircraft in such a manner that the formation could keep station without straggling. He had very sharp eyesight indeed and was constantly reporting enemy aircraft and manoeuvring the wing into a favourable attack position generally before others had picked up the target. Equally important, he led the wing in such a way that it was least vulnerable to being 'bounced' – eg: keeping the sun in the right place to the formation as far as possible, not turning down sun and having the squadrons stepped up or down sun as appropriate.

'He was a great tactician as well, and during bomber cover missions for example, would not be drawn away by the enemy but might despatch a section, flight or squadron to deal with targets of opportunity. He would always endeavour to give cover to such detached units. We normally maintained R/T silence until enemy aircraft were spotted or we were in the target area. Sailor provided almost a running commentary on the enemy aircraft situation and positions along with pretty forthright instructions to any squadron, flight or body out of position in the formation. As his Number Two one was, of course, sticking to him like glue whilst at the same time weaving like mad and endlessly scanning above and behind.

'Once Sailor engaged the enemy things were different – he was a most aggressive fighter, hard on his aeroplane (there was then only one place for the throttle at this stage – fully forward) and both hands on the control column! He was very strong as well, with a high "G" threshold. It was a private dog-fight for a Number Two to stay with him – he fairly tore into the enemy and I more than once recall pieces of Me109 flying past being the first indication of action, he was very quick in attack. He was an incredible marksman and a 'snap' shooter, miserly with his ammunition and never firing out of range – it was really hard work staying with him but very informative.

'On the ground his briefings before and debriefings after a 'show' were always meticulous. He particularly ensured lessons were driven home during debriefings in a fair but pointed manner and squadron, flight or section leaders (or individuals) seldom made the same mistake twice – it was good for us all and some lived longer for it. Those of us privileged to fly as his Number Two of course deemed it a great honour – we learned so much so quickly and it stood us in good stead in later days. It was, I suppose, a time for learning quickly and how better?

'He was a most human person with concern for his junior pilots but I seem to recall a certain ruthlessness – but little room for apparent sentiment existed in those days. He was relaxed and sociable off duty but I do not remember him as a great "party" man – he was solid and reliable and somewhat older in relative terms. He was a family man and lived out of the mess with his wife near Biggin – I recall swimming parties at his house even though I was a very junior person and the Wingco was a formidable person in those days.'

*

During the evening of 25 June, 12 Blenheims were the centre of Circus 23. Malan led his wing out for the second time that day, again as Target Support Wing, crossing the hostile coast at Gravelines between 25,000 and 29,000 feet, then gradually losing height towards St Omer-Longuenesse airfield – the target. No enemy aircraft were seen on the outward journey but on the return, again flying in fours, the wing saw several 109s and Sergeant Payne (W3308) of 92 Squadron destroyed one.

Circus 24 took place the following morning, 24 Blenheims raiding Comines Power Station. Malan's wing was once again Target Support Wing. Heading out over the green-grey waters of the English Channel, they crossed the French coast near Mardyck at 11.35 a.m. between 17,000 and 20,000 feet. Malan called 609 and ordered them to remain over Gravelines but later when 74 and 92 flew back from the target area, 609 were too high to engage the 109s which Malan found, as ever waiting for them.

Malan attacked one, knocking pieces from it and forcing it down, pouring smoke, but was only able to claim it as probably destroyed. Jamie Rankin also got in amongst the 109s – 24 he counted – which came up from the direction of St Omer. He called Malan to say he'd seen them, and Malan gave him the OK to attack. Rankin peeled off with his section and attacked the leading vic of Messerschmitts. He blacked-out in the wing-over but came too with a 109E diving away to his left, heading towards the coast. Firing his cannons and machine-guns from 200 yards he was buffeted by the 109's slipstream. However, it began to leave a thin trail of glycol before it went down and over onto its back to dive inverted. Rankin followed it down, seeing it hit the ground 10–15 miles inland between Dunkirk and Gravelines.

Total victories for the Biggin Hill Wing on the 26th were three destroyed, one probable and two damaged for the loss of one pilot. The total for Circus 24 was nine destroyed, four probables plus eight damaged for the loss of three aircraft.

*

There was a break in the scoring action on 27 June, the first in a week of intense action but it only lasted the one day. Circus 26 was scheduled for the 28th. Yet a sweep was flown on the 27th, over St Omer, with disastrous results. One pilot lost was Sandy Sandman, who had flown as Malan's Number Two on several occasions. He went down to become a prisoner of war (W3120). Sergeant Hilken (W3252) also failed to return but the really bitter blow was the loss of John Mungo-Park, CO of Tiger (X4668). He had over a dozen victories. He had also been due to be Sailor's son's godfather. Paddy Treacy had also been named for this honour but his death prevented it. Now Mungo-Park was gone. Strangely enough he was shot down in the Spitfire flown by Sandman when the latter shot down his two 109s on 22 June.

On the 28th, the Biggin Hill Wing was part of the Target Support Wing, the target being the Comines Power Station. The wing on this occasion comprised 27 Spitfire Vs, nine from 609 including Malan, 12 from 92 and six from 74 Squadron. Malan led them off at 7.50 a.m., the Tangmere wing being assigned as Cover Wing. The Biggin boys crossed out over the North Foreland, with 609 leading of course; all three squadrons were flying in sections of three and four. 609 were at 20,000 feet, 92 three thousand feet higher with 74 as rear cover flying at 30,000 feet.

The target was reached at 8.35 p.m. right on time, and Malan circled widely. Below, the two dozen bombers pounded the target before curving round for the homeward trip. As Malan turned his Spitfires on a north-westerly course, he saw several formations of Messerschmitts in twos and threes, but they all avoided combat. More enemy aircraft were seen about ten miles inland from Gravelines, flying in a westerly direction at heights ranging from 12,000 to 24,000 feet.

Some of these made a diving attack just behind Calais on what appeared to be the rear aircraft of a 12 Group Wing. Malan led 609 and 92 Squadrons down to the attack, picking out four Messerschmitts flying at 8,000 feet, leaving 74 Squadron to protect their backs, at 25,000 feet. The Germans saw them coming and by diving to one side and then climbing rapidly, managed to stay out of the Spitfires' range. Then one Messerschmitt dived right under Malan's machine to attack a group of Spitfires. His windscreen frosted up during his dive on the German fighter and when it cleared, he found himself mixed up with a number of Spitfires and two Me109Es. He latched onto a 109 and dived towards Calais and, while doing so, avoided the attentions of two other 109s who made menacing approaches on him from the left.

Malan eventually closed with his adversary when down to 8,000 feet, mid-way between Gravelines and Calais. The 109 went down to attack a Spitfire from its rear. Malan overtook it, opened fire from 200 yards with just his machine-guns, closing to 150 yards and then let go with his cannons and his .303s, firing a three-second burst. The Messerschmitt rolled slowly to starboard, clouds of black smoke and flame issuing from its engine. The 109 rolled again slowly before its nose dropped. Malan felt certain that the German pilot was either killed or badly wounded from the way his aeroplane went down.

Seeing more 109s diving from the direction of Calais, Malan climbed to 5,000 feet and cleared the area. He had fired 44 cannon shells and 240 rounds of .303. It is of interest to see how Sailor Malan on this occasion used his fire-power. There was a rocker-button on the control column which controlled the firing of the guns. If a pilot pressed the top part it fired the cannons, press the bottom part and just the machine-guns fired. Press the whole of the button and both armament fired. What did the successful air fighter usually do? Bob Stanford Tuck told me:

> Usually, if you were in a good position you'd just zonk off your cannon because that was your hardest hitting weapon. But if you weren't and you wanted to get as much ammunition at it (the enemy aircraft) as you could, and it was an unpleasant shot, you'd use the lot, cannon and machine-guns. In other words create a maximum lethal density by firing the lot.

The only pilot in 609 Squadron who had the opportunity on this mission to fire was Flight Lieutenant Paul Richey DFC. He climbed after a 109 which he saw turn onto the tail of a Spitfire, but, by turning tightly, Richey got onto the tail of the 109 as it began to fly a weaving dive inland. Richey fired three bursts of cannon from 250 yards before other 109s forced him to break off the action. Paul Richey had flown with 1 Squadron in France in May 1940, where he had won his DFC, gaining several victories before being wounded. His well known book *Fighter Pilot* is a classic of the Battle of France. Richey was also Mike Robinson's brother-in-law.

Rankin's 92 Squadron tried to engage four Me109Es but they dived away inland. On the way out the wing experienced heavy flak from the woods and sand dunes east of Calais.

The next day, the 29th, Malan listened to the radio at 1 p.m., hearing his pre-recorded BBC broadcast about some of the 'adventures' he had been involved in while leading his wing.

*

Circus 27 to Lens was flown on the last day of June. It was another power station, this time the one at Pont-à-Vendin. There were 18 Blenheims, ten from 18 Squadron, plus eight from 139 Squadron. Biggin Hill, with the pilots of the Hornchurch Wing, were the Target Support Wings. Rendezvous with the bombers and escort wing was made ten miles east of North Foreland.

On the approach to the target, Malan spotted a vic of eight to ten Me109Es which were making a gentle right-hand turn at

12,000 feet, about 5,000 feet below him. Giving the order to attack, he dived with 609 Squadron and two sections of 74 Squadron, straight towards the 109s, leaving the third section of 74 to cover them. All the Messerschmitts straightened into a fast dive with the exception of two which were flying on the extreme right. Closing in on one of these, Malan opened up with a continuous fire from about 350 to 100 yards, using both cannon and machine-guns. As he pulled away, the 109 was completely obscured by black smoke pouring back from its shattered engine which was well on fire. Its fall was confirmed by his Number Two, Sergeant W.G. Lockhart.

Robinson's squadron got into a fight with 109s; Paul Richey destroyed a 109E and shared another probably destroyed with Flying Officer Roger Malengreau (Belgian). Malan had led this attack also, and Richey picked out four 109Fs he saw to his left and above. He led his Yellow Section to attack them, destroying one. Richey and Yellow Three – Malengreau – joined up with 74 Squadron, who were flying north-west at 16,000 feet and climbing. Then six Me109Fs appeared above and behind, diving in open formation. Richey turned and in the heat of the moment he forgot 74's call-sign and was only able to swear at them for flying straight on! He and Malengreau attacked a 109E which was credited as a probable.

Squadron Leader Robinson (W3238) damaged a 109E. He saw these 109s well above the wing and in most cases they seemed quite content just to be a menace rather than to attack the Spitfires, but obviously ready to pick-off any stragglers. Robinson then saw two 109s making a pass at 74 Squadron and fired a burst at one which he then chased but he could not close the range. He then attacked a group of five 109s, sending one down streaming coolant and brown smoke. Two other 109s were destroyed by 609, one by Sergeant David Hughes-Rees soon to win the DFM. Malan and the wing landed back at Biggin shortly after 7 p.m.

It was at this time that Mike Robinson received the DSO, while Paul Richey was awarded a bar to his DFC. Jamie Rankin received the DFC while his two flight commanders, Alan Wright and T.S. 'Wimpy' Wade collected a bar to a DFC and the DFC respectively.

Sailor Malan's victory on the 30th brought his total victories for the month of June to ten destroyed plus one probable. Air Ministry issued a list of their top fighter 'aces' as at 30 June 1941. Malan's name was at the top with 29 victories. Bob Stanford Tuck was second with 26 while Ginger Lacey was third with 23. On the previous list issued at the end of March 1941, Lacey had been top with 23, Tuck fourth with 22, Malan sixth with 20. The leader of the Tangmere wing, Douglas Bader, was shown as having 15½ kills as at 30 June.

Another list issued gave results of operations by Malan's Biggin Hill wing as:

74 Squadron – 6½ destroyed, 8 probables
92 Squadron – 25 destroyed, 7 probables
609 Squadron – 14 destroyed, 5 probables

Malan's firm friend Bob Tuck was aware of the closeness of their respective scores. For some time they had been jostling for top position. It was not a race but it was obviously apparent that they were both top men in air combat. Tuck recalls Malan in these words:

> He was a very experienced fighter pilot. He knew his job inside out. He was a damn good fighter leader in the air, jolly good with discipline, and yet a very pleasant personality, but strong with it. He could shoot straight, flew well and was experienced, which was what it was all about. Sailor was quieter and more thoughtful but would have as much to say as anyone else when it mattered.

On the last day of June, 74 Squadron welcomed a new 'Boss' to replace Mungo-Park. For Malan it was the face of an old friend – 'Charlie' Meares. Squadron Leader Stanley T. Meares came in from 611 Squadron. He had been with Malan in 74 Squadron before the war and been a member of Malan's Flight in the Sassoon Trophy success. It was always reassuring to have old and trusted friends about. In the deadly game of air fighting one's life often depends on having someone you could trust behind you.

With men such as Meares, Rankin, Robinson, Richey, Wright, Bisdee and others, Malan's wing was a formidable team in the air war over Northern France in 1941.

CHAPTER EIGHT

Action over France

Wing Commander A.G. 'Sailor' Malan was beginning to tire. He had been a front line fighter pilot in peace and war for four and a half years, and for the last thirteen months he had been on operations with hardly a break. He had also been in constant command of flight, squadron or wing, which added tremendous additional responsibilities to his foremost job of getting to grips with the German Luftwaffe.

As July 1941 began there was not the slightest suggestion of any let-up in the almost daily round of offensive operations over Northern France. It was obvious now that the number of Luftwaffe fighters the RAF were meeting when they were enticed into the air was much reduced. Often, too, those the RAF pilots saw were reluctant to join combat unless they were in a very favourable position or they could pick off any stragglers. Since the German invasion of Russia, only two Luftwaffe *Jagdgeschwader* were left in France. These were JG2 'Richthofen' and JG26 'Schlageter'. JG2 generally operated to the west of the River Seine, in Normandy and over Cherbourg while JG26 flew east of the Seine over Northern France and Belgium.

However, there is no suggestion that the RAF were having it any easier since the invasion. JG2 and JG26 were full of experienced fighter pilots – and they were still a confident and victorious group of young men who, despite setbacks over England in 1940, were still very dangerous adversaries. And of course, the RAF were now flying over enemy-held territory where the Luftwaffe held most of the advantages. It was now the Spitfires that had often to fight their way home from France and across the Channel with drying fuel tanks.

Circus 29, flown on 2 July, was certainly no easier for the pilots of the Biggin Hill wing. The target was Lille, but it was the secondary target that was bombed, the airfield at Merville. Twelve

Blenheims comprised the bomber force, six each from 21 and 226 Squadrons. Malan's wing was the escort for the Cover Wing.

Malan led his three squadrons off at five minutes to mid-day, climbing to the familiar North Foreland rendezvous point, 74 leading, 92 in the middle and 609 on top. The wing was scheduled to leave North Foreland at 12.15 p.m. and actually turned onto a course of 120 degrees at 12.14. Two minutes after leaving the area, Malan, with the leading squadron at 19,000 feet, received a radio message from the Controller – 'Your friends are five minutes early.' Malan and 74 surged ahead but 92 Squadron (code name/ call-sign 'Garrick') and 609 (code call-sign 'Beauty'), misunderstood the message as 'They' were five minutes early. In consequence both squadrons throttled back and dropped behind.

Malan carried on and as he and 74 approached Dunkirk he received another message – 'Your friends are seven minutes early, proceed with all possible speed.' As Malan crossed the French coast he attempted to contact 92 and 609 to try and discover their present location. Robinson confirmed he could see Rankin's squadron but Malan himself was unable to establish any sensible communication between him and 92 Squadron.

Flying into France, Malan caught sight of the main formation and flew up to it about five miles inland. He carried on and when about five or six miles from the target, he saw some Me109s in small groups of twos and fours at 25,000 feet to the left. These 109s attempted to deliver attacks with their usual tactic of diving and then levelling out behind. It was then that Malan saw 92 Squadron above and behind and also the Tangmere squadrons flying some five miles off to the right and above, coming in on a converging course.

The German's attack appeared to be directed against 74 Squadron and a general melee began, but by getting the Tigers to turn in sections towards each attack, a certain amount of success was attained. Malan himself fired at one Messerschmitt from the beam as it went flashing past but he missed, but then followed another one down through some Spitfires from the Escort Wing and saw it off.

Sailor was still in contact with the main formation and the dog-fight continued to rage till well past the target area. Then two

109Es dived through the formation, before pulling up into a climb. Malan followed with his section and just as the leading Messerschmitt was nearing its stalling point, Malan nailed it with a burst of cannon fire from about 150 yards. There was a large explosion in the 109's fuselage and its nose went down vertically. Malan followed for about 8,000 feet; the German fighter was streaming vast quantities of petrol. Sergeant Lockhart (W3217), on Malan's wing, continued firing at the Messerschmitt and was given half credit for the kill. Lockhart had seen the 109s when Malan gave the Tally-Ho call and saw Malan open fire on one of them just before he opened fire on the same machine. He let go a two-second burst in the middle of which the 109 spun over onto its back and after the explosion went down vertically. As he watched it go down the black smoke turned to white. He lost sight of it at 7,000 feet. Charlie Meares saw the 109 hit, mainly in the front part of the cockpit and confirmed its fall to earth.

Regrouping, Malan found he had lost sight of the bombers. He collected together five of the Tigers and several of the Tangmere wing and got them into formation. His compass, following several aileron turns during the dog-fight, was spinning wildly – so which way to fly? Flying on judgement alone and being shadowed by two Me109s on the left some 4,000 feet above, with other 109s following in the distance, the Spitfires made for the coast. In order to prevent the two 109s reporting the Spitfire's course and position, Malan went for them, forcing them to dive away although lack of fuel prevented any real pursuit.

He eventually led the remaining Spitfires out at Etaples, near Boulogne, landing at Hawkinge with just five gallons of petrol left in his fuel tanks. Back at base Malan discovered that 92 Squqdron had also been in action; Don Kingaby claimed two 109s, and Sergeant Lloyd and Sergeant Pietrasiak each put 109s into the sea. These brought 92's total kills to 165.

*

July was proving just as hectic as June had been. Circus 30 was put on for mid-morning of the 3rd; six Blenheims of 139 Squadron attacked the marshalling yards at Hazebrouck. Malan's three squadrons would again be the Target Support Wing. It was a beautiful sunny day with clear skies. Although there were only six bombers, they had 16 fighter squadrons in the air, all taking some part in the operation in order to defend them.

Reaching the St Omer-Hazebrouck area, Malan broke the wing up into fours. Several Me109Fs could be seen in small groups but they did not seem keen to mix-it with the RAF fighters. It appeared in the main that the 109 pilot's intentions was to lure the Spitfire sections towards a large concentration of 109s south of St Omer. Suddenly Malan pulled round after some 109s, catching them by surprise. He got in a burst at a pair of them from 300 yards, and Charlie Meares saw one Messerschmitt begin to leave a trail of heavy smoke as it dived gently away to the left.

Malan turned northwards with his section and found several pairs of 109Fs at around 23,000 feet which were menacing a group of Spitfires 2,000 feet below them. A series of dog-fights broke out, and Malan reported later of the 109's extreme manoeuvrability. However, this was still below the performance of his Spitfires.

Splitting his section into two pairs he and his Number Two, who was none other than Biggin Hill's Station Commander, Group Captain P.R. 'Dickie' Barwell DFC, went after a couple of 109s. Malan turned wildly with them, and when one seemed to lose control for an instant, it was long enough for the deadly South African to take full advantage of his adversary's mistake. The 109 flopped onto its side and lost flying speed, appearing an almost stationary target as Malan jabbed the gun-button. At full deflection his cannon and machine-guns splattered the German single-seater. The 109 straightened up and went into a fast dive; the starboard undercarriage leg flopped down fully extended, while the port leg began to wave about loosely. Malan closed in for the kill and emptied his cannons from 150 yards, nearly colliding with the German. As his cannons stopped pumping, his machine-guns continued rattling. Then he had to break away to avoid hitting the 109, but as he did so, saw its port oleo leg rip away into the slipstream. Malan's break-off was so violent that he passed out for a moment and coming to could not see the 109 anywhere.

Rankin's squadron got into a brief battle, and Sergeant Don

Kingaby hit one Messerschmitt which he claimed as probably destroyed. 609 were approaching the target area when a lone 109 was seen 5,000 feet below them. Robinson was leading his men in loose fours and they were just north of Hazebrouck. Seeing the 109, Robinson half rolled and led Red Section down behind it. The German pilot took no evasive action, being totally unaware of the approaching danger. Closing right in to just 50 yards Robinson opened fire with cannons and machine-guns from dead astern. The 109 blew up, and its remains burned in a field after it smashed into the ground.

On the afternoon of the 3rd, six Blenheims from 18 Squadron returned to the same target for Circus 31. Malan's wing was again in evidence and Mike Robinson scored again. Over St Omer he spotted ten Me109s flying west at 3,000 feet higher than his own squadron and about one mile ahead. At the same moment he saw two Me109s 3,000 feet below, over Hazebrouck itself. He led Red Section down, but the Number Two of the Messerschmitts dived away. Robinson closed with the leading 109 who now pulled into a tight right-hand turn, but Robinson found no trouble in turning inside it. He fired a full deflection but the 109 climbed suddenly, then dived. Robinson followed, firing, and the 109 slowed into a near vertical climb, and Robinson had to break to avoid hitting it. The 109 flopped over to dive inverted, pouring out smoke. Sergeant Hughes-Rees watched it go, seeing its wings pull off just before it hit the ground.

Meanwhile the second 109 was fired at by Pilot Officer I. Du Monceau who saw it leave a trail of white smoke. Paul Richey, leading Yellow Section, saw three 109s below but his attack was thwarted by his Number Two who dived on a different 109, taking Yellow Three with him. Yellow Three had then to break off when attacked by a 109 with its sides painted red from its nose to the cockpit area.

During these combats on the 3rd, Hauptmann Wilhelm Balthasar, *Kommodore* of JG2 was killed. He had over 40 victories and the Knight's Cross with Oak Leaves. It was reported that he had a wing of his Messerschmitt shot off but whether this was Robinson's victory is not known.

The majority of the 109s encountered on this day, however,

made few moves to engage the RAF fighters, although 92 Squadron did lose one Spitfire (W4476).

On the following afternoon it was 21 Squadron's turn to provide six of their Blenheims for Circus 32 – the target being the Kuhlmann Works and Power Station at Bethune, while 12 Blenheims from 16 Group bombed the German air base at Abbeville. The Biggin Hill wing was assigned as Escort Cover. It was a bright, warm day. Take-off came at 2 pm and rendezvous with the bombers was made above Southend.

Sailor Malan, leading yet again, crossed the French coast ahead of his men at Gravelines, at 18,000 feet. Cloud began to build up which later developed to 10/10th at some levels inland. Then, somewhere in the vicinity of the target area, just as the bombers were starting to turn, several Me109Es which had been following above and behind, just waiting for the right moment to pounce, began a series of attacks by diving underneath the fighter cover in order to go for the bombers from dead astern.

Malan saw the first 109 coming and found that he was able to catch it up before it could reach the Blenheims. He opened fire with just his machine-guns as the 109 was almost upon the rear vic of bombers. Malan pulled away to avoid flying into the bombers himself but was sure he had damaged the Messerschmitt.

The South African then re-climbed to his main formation which was flying on the left flank of the bombers and at once saw two Me109Es which had lost most of their diving speed and were trying to climb away. He attacked one of these from 200 yards with both cannon and machine-gun fire which produced heavy black smoke. Sergeant Adolf Pietrasiak, a Polish pilot with 92 Squadron, (in W3245) also saw these 109s and made an attack at the same moment as the Wing Leader. He saw his own and Malan's fire hit the German aeroplane and watched as it went down in flames and belching an oily smoke plume. Shortly afterwards Pietrasiak saw a 109F diving at the Blenheims but it zoomed up before he could get his sights on. Waiting for a better chance, the Polish flyer bided his time for the same 109 pilot to try again, which he did. This time, as the German pilot zoomed up vertically, Pietrasiak attacked to fire a three-second burst with his machine-

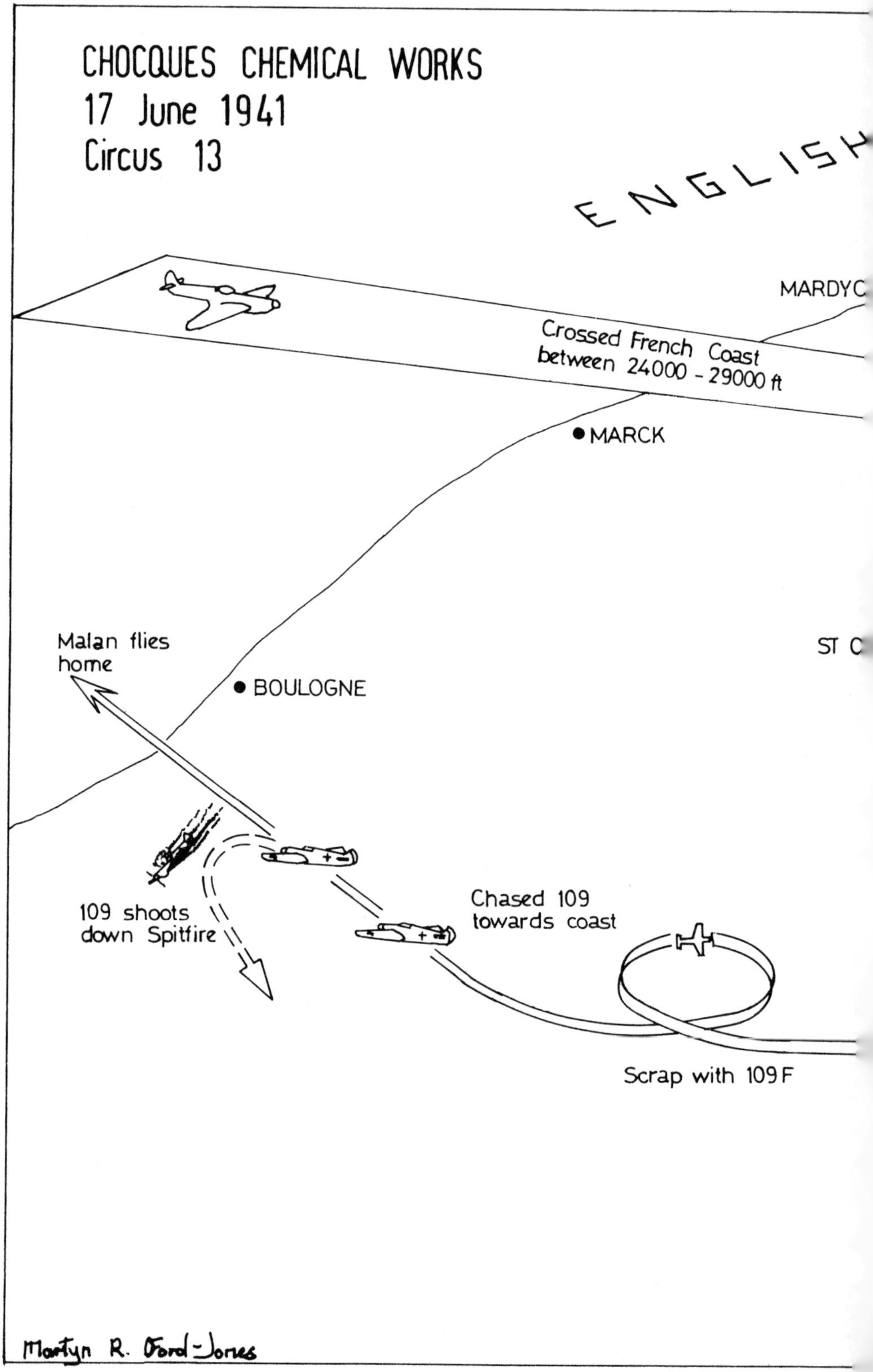
CHOCQUES CHEMICAL WORKS
17 June 1941
Circus 13
ENGLISH
MARDYC
Crossed French Coast between 24000 - 29000 ft
MARCK
ST C
Malan flies home
BOULOGNE
109 shoots down Spitfire
Chased 109 towards coast
Scrap with 109F
Martyn R. Ford-Jones

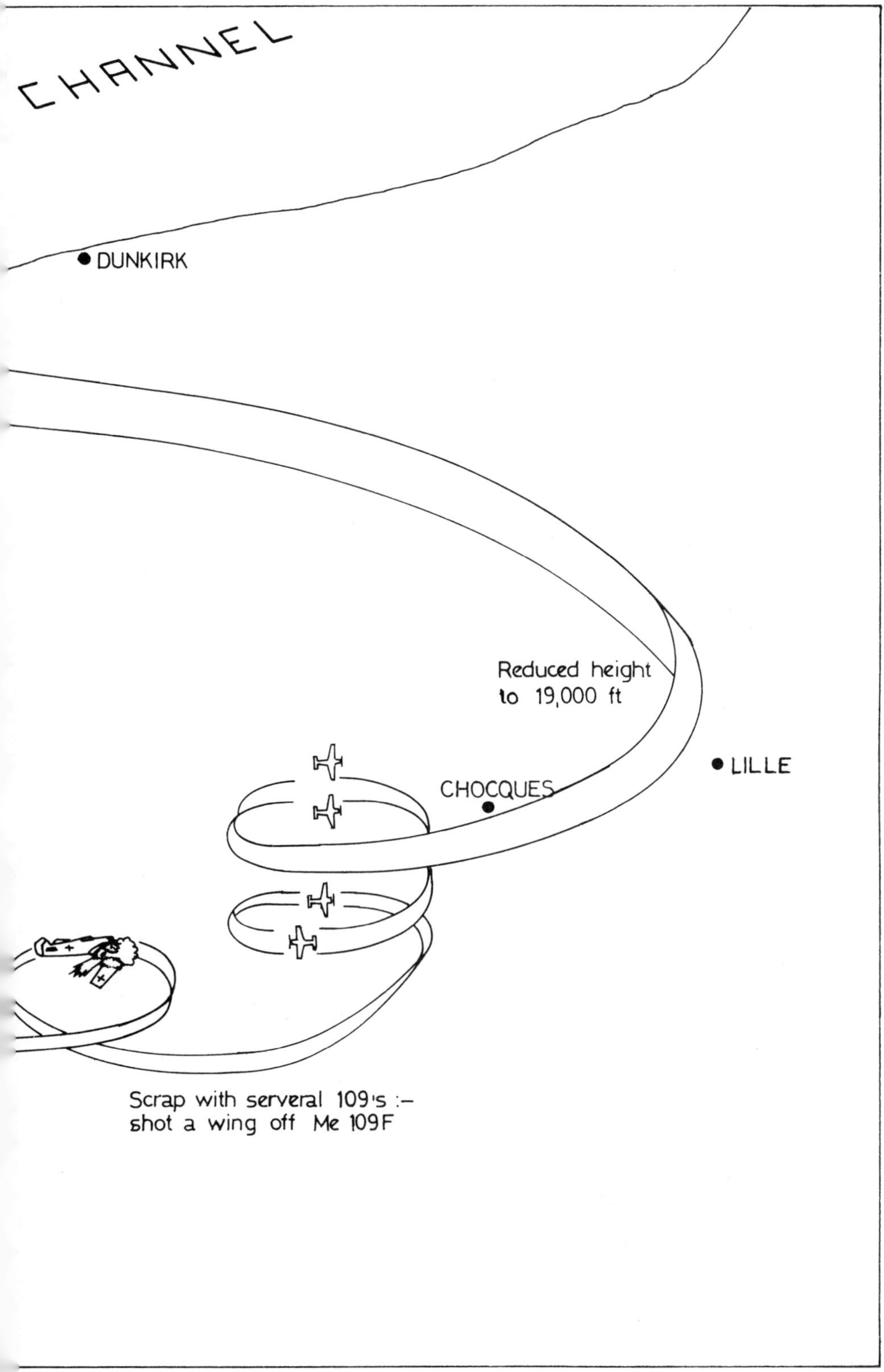
CHANNEL
DUNKIRK
Reduced height
to 19,000 ft
LILLE
CHOCQUES
Scrap with serveral 109's :–
shot a wing off Me 109F

guns. The 109 burst into flames and went down pouring out smoke.[1]

Meantime, another 109 appeared underneath Malan's port wing, flying slowly in the same direction as himself. Malan simply pushed his 'stick' forward as the 109 pulled ahead, lined up on the Messerschmitt and fired. Two seconds of cannon and machine-gun fire and the 109 exploded and fell earthwards on its side, streaming smoke and flames.

Group Captain Dickie Barwell, again flying as Number Two to Sailor, saw his leader busy with the 109 as another passed under him, and then begin to climb away to the right. Barwell turned after it to deliver a quarter deflection attack. The Messerschmitt trailed smoke and petrol vapour as it nosed over into a dive. Malan saw it going down and was able to confirm it as a probable. Malan and Barwell reformed and together they went into a steep turn to try and reform the wing which had drawn some distance ahead. Giving chase, they found themselves beneath 20 to 30 Me109s which were flying in the same direction as the main Circus formation although making no attempt to attack. Yet some did decide to attack the two senior Biggin Hill pilots. They both parried these attacks and more 109s joined in. As Malan attacked one 109, Barwell saw yet another right behind, diving towards them and about to open fire. With no time even to shout a warning to Malan, Barwell hauled his Spitfire round to face the 109 which promptly turned and dived away.

Now separated, Barwell, down to 8,000 feet, started to weave his way home but on the way he was attacked by 109s on several occasions, firstly by just single fighters, later by as many as five. All he was able to do was turn quickly towards the attacked and open fire when he got in range, although he was unable to see any results of his fire. He was eventually forced down to 5,000 feet when between Běthune and Gravelines but was able to fly into cloud and escape; another 109 made a pass on him when he emerged over the Channel when he was only five miles off Dover.

Once separated from Barwell, Malan also escaped in the

(1)Adolf Pietrasiak later won the DFM and was commissioned. He then flew with 308 Polish Squadron but was lost on 29 November 1943, when his Spitfire IX (MA584) developed engine trouble and fell into the Channel. He had previously flown in Poland in September 1939.

clouds, dodging several Messerschmitts by popping out of the top and bottom alternatively, and then altering course. Eventually he emerged from the top of the clouds over St Omer just as a 109E did the same on a course at right angles to his own. Sailor winged over towards it, fired a two-second burst from 200 yards and saw cannon strikes but nothing more. He saw no more as he completed a wing-over and dropped back into the cloud inverted.

Flying into clear air, he made for Gravelines, gaining height to 10,000 feet. As he headed for the coast in a shallow dive, an Me109F attacked him from behind but Malan had seen it and waited until the German was just 500 yards away then whipped his Spitfire round to face his enemy. The 109 pilot pulled away and Malan flew home.

Number 609 Squadron, flying as rear squadron, encountered 109s from just before the targe area onwards. Paul Richey damaged two and the CO, Mike Robinson, knocked pieces from another. Robinson saw four 109s diving down through his formation and fired one full deflection shot at all four as they passed through his sights, one of which streamed smoke. Yet they didn't get away scot-free. Flying Officer Keith Ogilvie DFC, the veteran Canadian, was shot down and taken prisoner. Six weeks later Lord 'Haw Haw' boradcast that Ogilvie was a captive. Ogilvie, Pilot Officer Vickie Ortmans and Malan had not missed a single Circus operation since they began on an intensive scale. Sadly, Ogilvie, who had fought in the Battle of Britain and who had gained several air victories, was about to start a well earned leave to Canada. As the squadron diarist remarked, Ogilvie, always cheerful and full of good humour, must have laughed a lot about how near he'd come to going home. Vicki Ortmons had been a pilot in the Belgian Air Force until the fall of his country. Making his way to England he flew with 229 Squadron in 1940, then in April 1941 was posted to the Belgian Flight in 609[1]

*

July 5 was another bright, clear, summer day. Circus 33 had one

[1]Ortmans became a prisoner on 21 October 1941. He died in an air crash after the war. His brother Christian also flew with 609 but he was lost on operations against the Japanese later in the war.

marked change in that, instead of the usual twin-engined Bristol Blenheim, three four-engined Short Stirling bombers comprised the raiding force. Each Stirling was able to carry a bomb load equal to at least six Blenheims. The target was Lille, Malan's wing flying as Target Support.

After a take-off around mid-day, Malan formed up his squadrons and led with 74 Squadron, arriving over Lille with his men, well ahead of the bombers as planned. As he circled the target area until the Stirlings arrived, small groups of Me109s could be seen building up some distance away and then when the bombers and Close Escort squadrons arrived, the Germans began diving from the sides. Malan parried one attack by two 109Es, opening fire from 300 yards. Strikes of both cannon and machine-gun sparkled on the Messerschmitt's fuselage before it dived away.

Squadron Leader Meares also fired on a 109, also claiming a damaged. Returning to coast near Gravelines at around 1.30 p.m., Meares saw about 20 'R' Boats –*Raumboots*, similar to the British motor launches, used for mine-sweeping and defensive coastal patrols. They were motoring in towards Dunkirk. Malan saw them too, counting 27 of them about five miles off the French coast. He radioed to Meares and both men pushed their 'sticks' forward and dived down to make a strafing run across them. Meares commenced firing from 1,000 yards, closing to 250 on the rear-most vessel; Malan raked the same boat moments later. As the two Spitfires pulled up and away, Meares looked back, and saw the boat facing the opposite direction to its original course and with black smoke coming from inside the cabin area.

It was a longer operation than usual, lasting 2½ hours instead of the more usual 1½ hours. Back at Biggin the ground personnel were waiting anxiously but the skies remained ominously silent and empty. When the adjutant and intelligence officer of 609 Squadron had decided that all had been shot down, news came that the wing was queuing up to get down at Manston and Detling with rapidly drying fuel tanks.

*

It was not often that Sailor Malan allowed his Spitfire to sustain damage in air combat. Indeed it was a mark of his extreme prowess that this rarely happened. But it happened on 6 July, a sign perhaps that Malan was in urgent need of a rest from operations.

In the morning of the 6th, 74 and 609 escorted three Blenheims on a Roadstead operation, attacking shipping reported north-east of Gravelines. Nothing was found, however.

Circus 35 was arranged for the afternoon – target Lille once again. Stirling bombers, this time six in number, were again to be the attacking force with Malan's wing down as Close Escort. Once again Malan broke his men up into easy to handle sections following their arrival over the killing ground of Northern France. His own section comprised Dickie Barwell as his Number Two, Meares and Sergeant Lockhart as his Three and Four.

As the main force reached mid-way between Lille and St Omer, Malan and his Section, flying at 20,000 feet, saw several Me109Fs beginning to overtake from behind. The four Spitfires broke into independent pairs, Meares and Lockhart turning to the left just before he and Barwell did likewise. There followed a series of dog-fights; both pairs became separated and then all four men were split-up and fighting alone. Alone over France even without enemy fighters about was a dangerous place to be, now it was more so. Time to get out!

At the first opportunity Malan dived away in a series of half-rolls and aileron turns until his airspeed indicator registered 400 mph. He pulled out at 5,000 feet and straightened out, heading towards Gravelines. He was about one mile from the coast with his nose down when, looking below, he saw a shadow move across his own shadow, flying south-west along the coast. Looking about him he quickly saw an Me109F and with good speed on his side, he turned onto its tail and, like the hunter he had become, quickly closed in for the kill. The 109 was flying at 500 feet and Malan, after a quick scan of the sky, got to within 400 yards, where he opened fire with a four-second burst from all his guns, as he closed right in to 50 yards. The 109 was enveloped in smoke, went into four very rapid flick spins and crunched into the ground.

Instantly his Spitfire took hits. Three bullets grazed his flying helmet then ricocheted off the bullet-proof windscreen and into the instrument panel. Malan instinctively pulled his Spitfire into a steep turn to the left and looking over his shoulder saw six Me109Fs screaming down towards him. Malan did a rapid wing-over and chased by the six Messerschmitts, dived to just 50 feet above the ground, shooting out over the coast at Gravelines, jinking his machine violently while receiving a good deal of attention from every conceivable type of light flak and even coastal gun batteries. Sailor took avoiding action by watching the fall of the bullets and shells as they churned-up the sea ahead and about him.

Having lost his six initial adversaries, Malan steered a course for the English coast through a sea haze but was repeatedly attacked by a number of 109Es which forced him to constantly turn towards them, forcing them to pull up. In Malan's opinion (and he would have known) these Germans were very bad shots for, as he admitted, they surprised him from behind on several occasions and the first he knew of their attacks (they were using the haze to cover their approaches) was when their bullets began striking the sea under one or other of his wings. He finally shook them off about mid-Channel but when near the North Goodwins he was attacked by another 109. Malan whipped round and very nearly got onto its tail but the German pilot's superior attacking speed enabled him to pull up and away into the sun. Malan landed back at base drenched in sweat.

Charlie Meares and Sergeant Lockhart saw two Me109Es dive on the bombers from above and behind. The two Spitfire pilots climbed – Meares attacked the rear one from 250 yards, seeing machine-gun strikes behind its engine which produced a stream of glycol. The 109 then dived steeply losing coolant. Sergeant Lockhart also claimed two Messerschmitts but failed to get home (W3717). It was a bad day for the Tigers for in addition to Lockhart, Sergeant Carter was lost, as was also the veteran Pilot Officer Bill Skinner DFM. He had shot down or shared in the destruction of more than a dozen German aircraft since the Dunkirk battles. He was also the longest serving Tiger. It was with much rejoicing that he was later known to be a prisoner of war.

Ninety-two Squadron lost one pilot, Sergeant Todd (W3331),

who was last seen leaving the formation over the French coast on the way out.

Malan was asked by a journalist one day, probably a day very like 6 July, when he returned from a scrap over France, 'How's Jerry these days?'

'Jerry,' replied Malan, accepting a cigarette, 'isn't the pilot he was.'

'What do you mean?' asked the journalist.

'I mean that in the Battle of Britain last year when the German pilots outnumbered us five to one, they were very cocky, very brave. But things have changed. Across the Channel now the numbers are a shade in our favour – and Jerry is not the pilot he was. I think the German pilots we are meeting now, whenever we can, are a contemptible lot.'

*

The following day, 74 Tiger Squadron flew its last operation as part of the Biggin Hill wing. It was an uneventful day. On the 8th the Tigers flew north to Acklington, never again to be associated with Biggin Hill from where it had flown so often and for so long. It was a sad parting for Sailor too. He had begun and been with the Tigers or in close association with it for such a long time it was like losing one's home or a close friend. After all he had only ever known one squadron. However, he was destined to be with his beloved Tigers again before the war finally came to an end.

*

With the leaving of 74 Squadron, Bobby Oxspring's 72 Squadron flew in to take its place. Squadron Leader Oxspring DFC was another experienced leader and former Battle of Britain pilot already with several victories to his name.

It was almost as if the break up of the original Biggin Hill wing was the beginning of the end of Sailor Malan's fantastic period of successful air combat. No one who knew him in the summer of 1941 would argue the point that he was physically if not mentally tired, by the high summer. Nearly fourteen months of continual operational flying had to leave its mark on any man and, of course, Malan was also nearing his 31st birthday. Increasingly he began to stand down as Wing Leader, his place being taken by Squadron Leader Mike Robinson who was deputy Wing Leader. Yet Malan's reward for his tremendous success during 1941 came with the award of a bar to his DSO. With this new award Malan became the first fighter pilot to have won two DSOs and two DFCs. The citation to this decoration recorded Malan's '. . . great courage and disdain of the enemy.' also his '. . . cool judgement, exceptional determination and ability.' These were just not empty words – these summed up Malan the fighter and tactician exactly. Little wonder that his old friend from 54 Squadron, Al Deere, with whom he would fly again, says of Malan:

> Malan was the best. A great shot and splendid Wing Leader. He had the maturity of years and experience behind him. Others were good but he was the greatest.

*

Mike Robinson was a natural to take over from Malan as Wing Leader. He was many people's idea of the typical fighter pilot. Good-looking, an impeccable background, well-off family, and a good shot and pilot. He had flown with 111 Squadron before the war, 87 Squadron in early 1940, then with 601 and 238 Squadrons in the Battle of Britain before being given command of 609 Squadron AAF in October 1940. By mid-1941 he had more than a dozen victories and a DFC.

He led the wing on 11 July on Circus 44. Dickie Barwell also went along and managed to destroy a 109 and knock pieces off another. Robinson himself proved on this operation that he too was a worthy successor to Malan. It might have been so easy for him to rush in with disastrous results but he had seen the Master, Malan, in action over several months and had learned well how to conduct himself and a fighter wing.

When about ten miles inland from the French coast and heading towards Cassel, Robinson spotted several 109s climbing up through some cumulus cloud about ten miles away. He felt certain the Messerschmitts were gaining height in order to attack the wing from up-sun off Gravelines, the wing's point of departure.

Robinson led the wing parallel to the French coast, 15 miles inland and intercepted the enemy at a superior height between Cassel and St Omer. He first saw ten 109s, in no set formation, but did not attack at once, being convinced there were more following in the clouds. As he waited ten more 109s appeared. Watching carefully he saw a third group of Messerschmitts fly into view. Now was the time to attack.

Down he went, selecting a 109F, hitting it with cannon and machine-guns from dead astern from 700 to 150 yards. The 109 pulled up, pouring smoke and its pilot jumped.

Robinson got another 109 the next day, seeing it spin right into the town of St Omer and two days later, the 14th, he got another one. The pilot of this one also baled out, right in front of him, and the 109's hood hit his Spitfire's tailplane. On the 19th he damaged a 109 which he thought must have been piloted by a German 'ace' for he saw many victory marks on the rudder of his aeroplane. Robinson commented that he was lucky the Germans did not add another mark on that day.

Sailor and Robinson received the Belgian Croix de Guerre on the 22nd, for their work and encouragement with the Belgian pilots who flew with 609 Squadron. For Malan it was one of several foreign decorations he received for his services, others being the Czechoslovakian Military Cross, the French Croix de Guerre with Palm as well as being made an Officer of the French Legion of Honour.

Malan was back in harness for Circus 60 on 23 July. The target for six Blenheims of 114 Squadron was the oil refinery at Mazingarbe. It was a fairly late operation, not starting till 7.40 in the evening, and a whole lot of things went wrong.

As Malan formed up his two squadrons, 72 and 609, two pilots dropped out with oxygen trouble and a third had an undercarriage problem and then joined up with another Spitfire formation. Group Captain Barwell flew, leading Yellow Section, but his R/T let him down and he said he was returning to base. His section misunderstood and instead of going on, followed Barwell home.

Flight Lieutenant Paul Richey thus found himself with just four aircraft, and then he too was plagued with a faulty radio and lost touch with the leading squadron. While orbiting south-east of the Nieppe Forest, they were attacked by five Me109s. Richey led his men into a rapid turn which foiled the German's attack and then two Me109Es appeared ahead of them. They turned away as the Spitfires attacked, but Richey's fire hit one 109 which shed bits and pieces, and then spewed back quantities of oil. His Number Two, and Malan, saw it go down and later 72 Squadron confirmed seeing the pilot take to his parachute.

Having crossed the Channel, Malan found a huge shelf of cloud which stretched down the Strait between 26,000 and 30,000 feet and he tried to lead his pilots round it. However, he discovered that this would take them north to Dunkirk so he headed instead for Calais.

Soon afterwards he saw 16 Me109s in the St Omer area and dived to attack. However, the 109s, as usual were not playing and dived away, so Malan climbed again. Several more 109s could be seen and these and the wing met, which resulted in the start of a general dog-fight. The battle continued all the way to the target and back to the coast.

During these series of air fights, Malan saw that several 109Fs had managed to get inside the main formation but that they were unable to turn tightly enough to open fire on the Spitfires. Seeing one Messerschmitt below, Malan half-rolled and dived towards it, firing all his guns. Strikes blasted holes on its port wing-root and fuselage, then the 109 rolled over and dived. A yell over the R/T warned Malan of more 109s coming down, so he had to break away rapidly.

Ninety-two Squadron, flying a supporting sweep, encountered several 109s, and Sergeant Hickman shot down one in flames. They lost Flying Officer McDougal (W3120) who was taken prisoner. The squadrons began landing back at about 10 p.m.

The next afternoon saw another big operation for the wing and the last in which Sailor Malan was to score against the Luftwaffe. He led 28 Spitfires away from Biggin Hill at one minute past two o'clock on Circus 61, nine Blenheims attacking the Hazebrouck marshalling yards. Rendezvous with the bombers was made over Manston, and Malan's squadrons stepped up from 19,000 to 25,000 feet.

The wing crossed the French coast in visual touch with the Blenheims and Close Escort squadrons. Long before the French coast was reached large formations of Me109s were seen and after crossing into France, Malan ordered the squadrons into sections and to attack. He ordered 609 to turn towards the main formation, 92 to go after a group of 109s bunching up to the left, while he sent 72 in a climb to the right to engage another bunch of 20 to 30 109s coming in from the direction of Le Touquet at 25,000 feet. However, these 109s climbed away from 72 Squadron.

Other 109s were more aggressive; individual dog-fights were attempted by the wing; all over the area Dunkirk-Hazebrouck-Gravelines but the 109s were only eager to dive away and escape a general fight. As always the 109s were only keen to mix-it if they could pounce on a straggler, most totally avoiding any direct combat.

Most of the Messerschmitts encountered were 109Fs but one or two 109Es were seen. They were all coloured grey, one being seen with a yellow rudder. Malan and 72 Squadron then turned towards a formation of 20 Me109s and went straight into the attack but the 109s immediately dived. Malan and one or two others fired at these but could make no claims. They then turned towards yet another gaggle but these too dived rapidly. One section of 72 Squadron were jumped by six 109s flying at 18,000 feet and Malan chased after two of them, opening fire from 200 to 150 yards. He hit one, seeing strikes and pieces knocked from it but it half-rolled and dived. Being in no position to watch or follow it he climbed back to 24,000 feet where he chased two more 109s but these also evaded combat.

Large formations of 109s were seen by 609 Squadron as soon as they crossed into French air space. Mike Robinson attacked one group of 12 Messerschmitts but these too refused to fight. He did get one burst at another 109, blasted a chunk from it before it went into a violent spin. Sergeant Tommy Rigler fired at a 109, felt sure he had missed, but then the German pilot baled out!

Four 109s received attentions from 92 Squadron and in a

scrap that followed, Flight Lieutenant Brian Kingcome fired at one but had to break off without seeing any results. Flight Lieutenant Thompson hit another, whose pilot escaped by parachute. Neville Duke was jumped by two 109s over Gravelines. They came from out of the sun and it took violent evasive action by Duke to get away, but they chased him for 15 miles out to sea before giving him best.

Sergeant Johnson fired at a 109, one of two that attacked him just south of Dunkirk, and it went down in a spin, pouring out smoke. A few moments later he fired at another 109 which flew right through his burst. The 109 rolled over, went down in a spiral dive, trailing smoke.

The total for the RAF that day was five destroyed, four probables and four damaged for the loss of four pilots. One of the losses was Sergeant Vinter of 92 Squadron who went down ten miles south of Calais.

This was the end of Sailor's scoring. He had come a long way since he'd first smelled cordite over the Dunkirk evacuation in May 1940. He is variously credited with 28, 32 or 35 victories between May 1940 and July 1941. 28 really refers to the citation to his DSO bar or to those he was credited with on his own. There were seven kills in which he participated making 35. Depending on how one decides to add up bits of kills one could make a total of 32. This author has always preferred to give credit to any pilot as 'X' aircraft destroyed or shared destroyed. Therefore, as far as this story is concerned I shall take 35 as being Malan's total, ie: 35 destroyed or shared destroyed. In addition he claimed seven probables (including one shared) plus a further 14 damaged. A total, if you like, of 56 victories!

A yardstick of a fighter pilot's prowess is not, however, just an impressive list of victories. There is much more to it than a pure score. One might also be able to prove that such and such a pilot did not really achieve a given number of kills. That is easy if one can accurately piece together losses from the opposite side and compare them with claims. However, what we are accepting here is the claims at the time and the victories officially credited to Sailor Malan.

It should also be remembered that even if some of his

confirmed victories could be proved to have got back home safely(!), one would by the same token, have to prove that some of those he thought only damaged did not in fact crash.

CHAPTER NINE

Rested

Sailor was taken off operations. He had just returned from a three day weekend break which he and his family spent in a Kent cottage. He had been able to relax at last and as he later admitted:

> It was queer – and salutory. I heard the birds for the first time and smelt flowers. I did a bit of rabbit shooting, looked at the scenery, heard people talking about ordinary things and suddenly I knew how really clapped-out I was.[1]

For some time he had been encouraged to take a rest. He had done more than his share – more than a good job. He had always refused, his reasoning hard to explain even to himself. He had continued thinking to himself, 'Just a few more shows before I go', and of course he had got a few more, then a few more still. His shooting eye had been in; he was in peak condition despite obviously tiring. It was always easier somehow to continue than to rest and then come back. As the war was to prove, many fighter pilots returning from a rest period following a successful period of operational flying, just seemed unable to get it together again and on occasions they were lost soon after their return to front line duty.

Yet now, in August 1941, Malan was finished and forced to take a well earned and well deserved rest. Squadron Leader Mike Robinson took over command of the wing and in August he received the DSO. Jamie Rankin received a bar to his DFC. Robinson, however, was also nearing a rest period and in fact only led the wing until the end of August when Rankin took over as Wing Leader. His place as CO of 92 Squadron was taken by Flight Lieutenant R.M. 'Dickie' Milne DFC who had recently joined 92. He had won his DFC in the Battle of Britain. Charlie Meares too

(1) *Sailor Malan* by Oliver Walker. Cassell & Co, 1953

left the wing, taking command of the American volunteers in 71 Eagle Squadron on 22 August. He was awarded a well earned DFC and was married shortly afterwards. Tragically he was killed in a mid-air collision on 15 November 1941 during a training flight from North Weald. So another friend had gone.

It could be argued that Malan was rested just in time, for there were several casualties at that time, of important and experienced air leaders. On 9 August the seemingly undestructible Douglas Bader, leader of the Tangmere wing, went down following a collision with a Messerschmitt 109 over France, to become a prisoner of war. Wing Commander E.N. Ryder DFC and bar, leader of the Kenley wing, was also taken prisoner and Johnny Peel DSO DFC, who took over this wing for the second time, was shot down for the second time during August 1941, fortunately being rescued from the Channel. Wing Commander John Gillan DFC and bar, leader of the North Weald wing was shot down and killed on 29 August and Wing Commander J.R. Kayll DFC had been taken prisoner in July when leading the Hornchurch wing. Even Bob Tuck went down to ground fire early the following year to be taken prisoner. The 'Old Guard' was thinning out – the Few were becoming fewer.

*

Sailor Malan's first job after leaving Biggin Hill was as Chief Flying Instructor at Number 58 Operational Training Unit, although he was only here for a very short time. Before this, however, on 10 August, his son Jonathan was christened. The problem of one of the child's godfathers had been solved following the deaths of Treacy and Mungo-Park. During a visit to Biggin by the Prime Minister, Winston Churchill, Malan had asked the great man if he would consider being a godparent. Churchill agreed. Unhappily on the actual day of the ceremony, Churchill was in mid-Atlantic on his way to a meeting with the President of the United States. However, he sent his brother, Major John Churchill, to act in his stead, the Christening taking place at Holy Trinity Church, Crockham Hill, Kent.

In the autumn, six flyers were chosen from the Royal Air

Johnathan's Christening. Sailor and his wife Lynda with Major John Churchill, who stood in for his brother Winston Churchill as Godfather.

(Left) Harry Broadhurst who led the fighter pilot contingent to America in the autumn of 1941, comprising himself, Malan and Bob Standford Tuck. ***(Right)*** Friend and fellow fighter pilot Bob Stanford Tuck.

Force, to travel to the United States of America to talk to American pilots, give a series of lectures and to try out various American aircraft. On this last item they would also advise the Americans on operational requirements of machines being built for use by the RAF. The six men chosen were three fighter and three bomber pilots. Malan was chosen as was his good friend Bob Stanford Tuck. The third air fighter was Group Captain Harry Broadhurst. Broadhurst was another pre-war pilot, who had recently not only commanded RAF Station Hornchurch but had been leading the Hornchurch Wing. He was older than Malan but this had not stopped him flying on operations and destroying a number of enemy aircraft, collecting a DSO and DFC for his achievements.

The three men from Bomber Command were Group Captain John N. Boothman, a former Schneider Trophy pilot who had lately been planning and taking part in raids on the Ruhr, and Wing Commander J.N.H. 'Charlie' Whitworth DFC who had flown 38 bombing missions. The third man was Wing Commander Hughie Edwards, holder of the Victoria Cross for his daring low level attack on Bremen in July 1941. He also held the DFC.

America was, of course, still neutral, so the six men had to sail across the Atlantic in civilian attire. They sailed in the *Louis Pasteur*, arriving in Halifax, Nova Scotia, on 25 October. Sir Harry Broadhurst recalls the trip as follows:

> There were three of us fighter pilots, and we sailed in the *Louis Pasteur* arriving in Halifax on 25 October 1941. On the 26th, we flew to Washington, where we had talks with the senior staff of the Army Air Corps and to arrange our programme. On the 28th we flew to New York, where we met the Press and on the 30th we flew to Dayton and from there we were at Wright and Patterson Fields where we inspected the latest American fighters and talked to the test pilots. On 1st November I flew a P38 Lightning and on the 2nd I flew a P38 and had a dummy combat with Tuck in a Spitfire at 28,000 feet and then a P43, again in combat with Tuck in a Spitfire at 25,000 feet. The next day more discussions with test pilots and the Commanding General, Kenny, and thence to

Washington where I had an interview with General Arnold who was Chief of the US Army Air Corps. On the 5th I had meetings with the senior RAF officers which included Bomber Harris. Then on the 7th we flew to Laketown Training School where we lectured the pupils and instructors and on the 9th we flew to Spartanburgh for the Army air manoeuvres and I was attached to Number 1 Pursuit Group. On 10 November I flew to Charlotte and talked to Support Command HQ and then flew back to Spartanburgh, where I dined with Colonel Eaker, later to command the American 8th Air Force in UK, who was commanding the Fighter Wing. Flew to Columbia on the 11th and gave talks to Numbers 27, 94 and 71 Squadrons and also the Wing HQ. On the 12th Tuck got lost in a P43 and forced landed safely. I flew to Columbia – picked him up and we went on to Jacksonville. On the 13th, lectured to pilots of the 49th Pursuit Group and on the 15th flew operations in P38s of the 1st Pursuit Group. On 16th left Spartanburgh, picked up Malan at Florence, and flew on to Washington for a conference on the 17th on the Army air manoeuvres with officers of the War Department. On the 18th flew to the Republic Works at Long Island and then I flew and assessed the prototype P47 Thunderbolt and had discussions with designers and test pilots. Flew to New York and gave a talk about the war to members of the Stock Exchange at Wall Street.

We now split up and the other two were taken to Hollywood for what I gathered later proved to be a bit of a thrash. I caught the night train for Montreal.

Broadhurst continued his tour in Canada, flying and lecturing and on 25 November took-off for Bermuda en-route for UK in a Catalina – the engines failed, and he crash-landed, being rescued by motor launch. He flew to Montreal again and on the 26th he left for Ganda in a Liberator, leaving for Prestwick shortly afterwards. On arriving home he found that the Americans were finally in the war. Broadhurst continues:

I have given you my itinerary, because Malan must have been

> with me most of the time, except when we were at the Army air manoeuvres when we were attached to different Pursuit Groups. The idea being, of course, to spread our know-how to as many American fighter pilots as possible in the time available. On the big occasions I obviously did most of the talking, because I was the senior member of the party, but when meeting pilots, Tuck and Malan would have been equally occupied.
>
> There must have been a report written when we returned to England, and I note from my diary that I had a meeting with the AOC 11 Group the day after I got back, with the C-in-C Fighter Command the following day, and with the Chief of the Air Staff a day or so after that.
>
> I suppose I could tell you the odd story about the trip to America, but I am afraid they wouldn't be suitable! – or printable!

Finally, Sir Harry sums up Malan: 'Along with many others I recognise him as a first-class fighter pilot and his contribution to the Battle of Britain, tactically, was of inestimable value to the success of that battle.'

*

On one particular occasion in America, Malan and Tuck flew on opposite sides during an air manoeuvre. Malan led the Blue forces, Tuck the Red. They carried out controlled attacks in the manoeuvre area; Malan's forces were flying P38s, Tuck's P43s, a forerunner to the P47 Thunderbolt. One thing Tuck used to do while in the United States was to take great delight in introducing Sailor to the various people they met. Making sure everyone could hear him he would say – 'May I present Adolf Gysbert Malan . . .' giving full emphasis to his first two names, especially the 'Adolf'. It infuriated Malan but he took it in good part. Tuck remembers:

> Several things we criticised. The Americans had aircraft which could fly high but they rarely went very high. Having come from the fighting over France and Europe, height was a basic

> principle. We had to get them up high and as we were leading them they had to do as we said. They were a good crowd of chaps, very keen but usually flew at 15,000 feet when we wanted them at 25,000 feet. They were not used to flying high and of course the aircraft handles differently at higher altitudes, but they picked it up very quickly.
>
> They had oxygen lungs on their chests. At height these would gradually form ice inside and had to be continually squeezed to crush the ice in order to keep it functioning.
>
> We had great fun in America, were feted and entertained. A pleasant trip if very busy. We were briefed most carefully before we went out there not under any circumstances to get into any discussion with anybody at any social function etc: because of the very heavy isolationist policy, so we avoided it like the plague.

At first Malan had some difficulty actually being allowed to get into the air. He had visited six or seven air bases, giving pilots talks on current fighter tactics as used in Europe. At the seventh base he virtually insisted on being allowed to fly one of the American aircraft and was given a Bell P39 Aircobra. The American commander had twelve of his pilots already airborne and radioed them that Malan was taking off to intercept them. Sailor had never before flown an Aircobra but having been shown the taps and tits etc took-off and was back in less than five minutes. In that period, as his camera-gun later proved, he had bounced and 'shot down' all twelve American aircraft. According to Sailor it had been easy as the Americans were still flying a type of formation the RAF had used as far back as 1939. It was recorded that Malan was rather depressed at what he had found in the States as far as formations and tactics used were concerned.

On the way home, the team returned to Canada where Malan and Tuck talked to the press although the press-men seemed more interested in the news of battles being fought in Libya than of the two pilots' impressions of their American tour. Tuck, like Broadhurst, remembers hearing about the Japanese attack on Pearl Harbor on the aeroplane home. Suddenly the team hoped that the Americans had listened carefully to what they had been saying to them over the last few weeks.

*

Upon his return from America, Malan was put in command of the Central Gunnery School at RAF Sutton Bridge. Here he spent the whole of 1942, but interspersed with visits to factories to show the flag to the workers, in addition to giving talks on air gunnery and fighter tactics to other training establishments. He also formulated and wrote his now famous *Ten Commandments* – his ten rules for air fighting. These pearls of wisdom from the mind and the pen of one such as Malan were much needed. They were printed and distributed to all fighter bases where they were usually pinned to notice boards in crew rooms, in briefing rooms and in training schools. His ten rules were:

> Generally speaking, tactics in air fighting are largely a matter of quick action and ordinary commonsense flying. The easiest way to sum it up in a few words is to state that, apart from keeping your eyes wide open and remaining fully alive and awake it is very largely governed by the compatabilities of your own aircraft in comparison with that flown by your opponent. For example, in the case of the Spitfire versus the Me109F, the former has superior manoeuverability, whereas the latter has a faster rate of climb. The result is that the Spitfire can afford to 'mix it' when attacking, whereas the Me109F, although it tends to retain the initiative because it can remain on top, cannot afford to press the attack home for long if the Spitfire goes into a turn. Obviously there are a lot of factors involved which must govern your action in combat – such as the height at which you are flying, the type of operation on which you are engaged, the size of your formation, etc:
>
> There are however, certain golden rules which should always be observed. Some are quite obvious whereas others require amplification. Here they are:
>
> (1) Wait till you see the 'whites of his eyes' before opening fire. Fire bursts of about one to two seconds and only

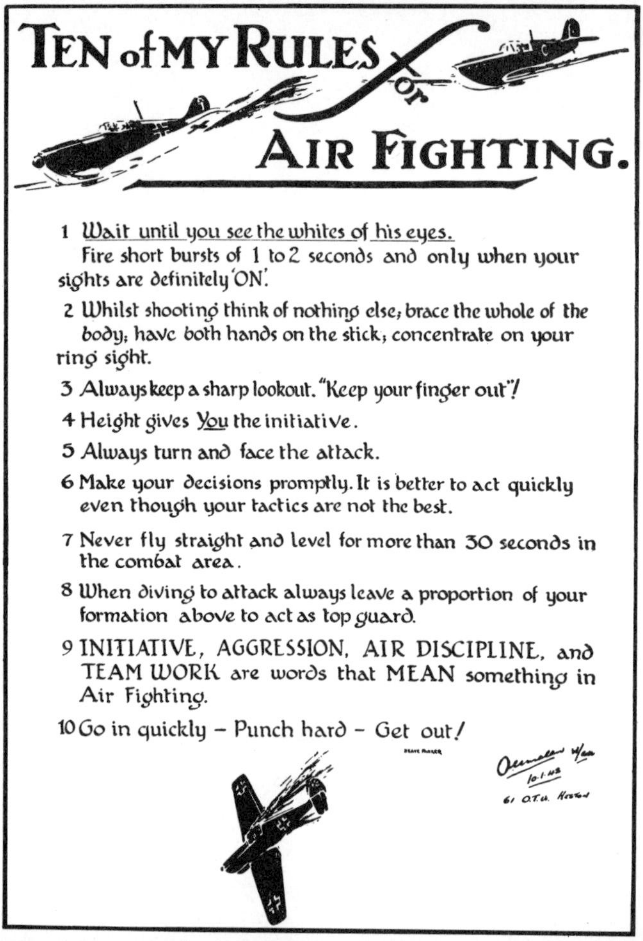

TEN of MY RULES for AIR FIGHTING.

1 Wait until you see the whites of his eyes.
Fire short bursts of 1 to 2 seconds and only when your sights are definitely 'ON'.

2 Whilst shooting think of nothing else; brace the whole of the body; have both hands on the stick; concentrate on your ring sight.

3 Always keep a sharp lookout. "Keep your finger out"!

4 Height gives You the initiative.

5 Always turn and face the attack.

6 Make your decisions promptly. It is better to act quickly even though your tactics are not the best.

7 Never fly straight and level for more than 30 seconds in the combat area.

8 When diving to attack always leave a proportion of your formation above to act as top guard.

9 INITIATIVE, AGGRESSION, AIR DISCIPLINE, and TEAM WORK are words that MEAN something in Air Fighting.

10 Go in quickly – Punch hard – Get out!

10.1.43
61 O.T.U. Heston

(Left) Sailor Malan's Ten Rules for Air Fighting.

(Below), Group Captain Sailor Malan, seated at his desk at Biggin Hill, 1943.

when your sights are definitely 'on'.

(2) Whilst shooting think of nothing else. Brace the whole body with feet firmly on the rudder pedals having both hands on the stick. Concentrate on your ring sight.

(3) Always keep a sharp look-out even when manoeuvering for and executing an attack and in particular immediately after breakaway. Many pilots are shot down during these three phases as a result of becoming too absorbed in their attack. Don't watch your 'flamer' go down except out of the corner of your eye.

(4) If you have thc advantage of height you automatically have the initiative.

(5) Always turn and face an attack. If attacked from a superior height wait until your opponent is well committed to his dive and within about 1,500 yards of you. Then turn suddenly towards him.

(6) Make your decisions promptly. It is better to act quickly even if your tactics may not be the best.

(7) Never fly straight and level for more than 30 seconds at any time whilst in the combat area.

(8) When diving to attack always leave a proportion of your formation above to act as top guard.

(9) INITIATIVE: AGGRESSION: AIR DISCIPLINE: TEAM WORK, are words that mean something in air fighting.

(10) Get in quickly – punch hard – get out!

The actual printed rule sheet was slightly abbreviated from the above but the main points were made. As we have seen in the previous chapters Malan always did just as his rules suggest. Almost always his first and many of his subsequent bursts for gunfire lasted for only two seconds. Rarely did he waste his ammunition. And from the amount of damage he inflicted with these small bursts his sights were definitely 'on'. He always faced an attack unless he was low on fuel or in a situation that really called for him to get away fast. During his period as Wing Leader he always left at least a section above to keep cover, although following the start of a major conflict everyone was eventually split-up.

*

On the second anniversary of the greatest day in the Battle of Britain, 15 September, a dinner was held to commemorate that victory. It was held at the Savoy Hotel in London and attended by a select group of veterans from the Battle. Malan was invited and so were other distinguished flyers from 1940, many of whom were still flying operationally. Al Deere was there, so were Max Aitken and Desmond Sheen; the diminutive Ian Gleed was there, so were Richard Hillary, Tony Bartley, Brian Kingcome, Douglas Watkins and Johnny Kent. So too was Lord Dowding, who had been Fighter Command's Commander in Chief in 1940.

Also on 15 September 1942, Malan made yet another broadcast, this time to his native South Africa. He spoke of both the day and night offensives being carried out by the RAF against Germany and concluded:

> Not all my fellow pilots in 1940-41 have lived to see us gain supremacy in the air. As I grieve their loss I consider the legacy they have left the entire Allied air force. It is for us to remember that those who died two years ago made possible the ascendancy we now have and will hold. Because of them I have been able to talk to you in South Africa, from Britain and to wish you 'Alles van die beste' – All the best.

*

Among the many pilots that passed through the gunnery courses at Sutton Bridge were men from all the Allied nations. British of course, also Australians, New Zealanders, Canadians, and South Africans. From Europe there were men from France, Belgium, Poland, Norway and Czechoslovakia. Even Americans attached from the USAAF or those others in the RAF not contemplating the move to their own air force.

Most of these were new boys, eager to get into action, but there were also courses for experienced pilots, joined together in order to pool their collective knowledge to help formulate new training methods and to see how best to pass on this knowledge to others. It was all very interesting and not a little rewarding but by the autumn of 1942 Sailor was agitating for an operational post. He had been away from the shooting war for well over a year and a spirit like Malan's could not be suppressed and kept away from the action for too long.

Finally he got his wish and he was soon to be in command of men from every nation. Just as he had helped to train such men, to give them some guidance of how to shoot down German aeroplanes, so now he could be with them 'down the sharp end'. He was promoted to Group Captain, which meant a fourth ring on his sleeve and 'scrambled egg' on the peak of his service cap. Yet far more came with it. He was back where the real action was. Back with the men who did the fighting, his kind of men. But even this was secondary to the actual job. One doubts if he could have asked for a better assignment, for he was given command of Royal Air Force Station Biggin Hill!

CHAPTER TEN

Station Commander – Biggin Hill

Group Captain A.G. 'Sailor' Malan DSO and bar, DFC and bar, took over as Station Commander RAF Biggin Hill on 1 January 1943. The posting received much coverage by both press and radio, Biggin Hill being described as Britain's number one fighter station. Malan, of course, was still Britain's number one fighter pilot.

There had been a great many changes since Sailor had last flown from 'Biggin on the Bump'. Dickie Barwell, the Station Commander when Malan had been leading the wing, and who had flown as Sailor's Number Two on several occasions, was dead. He had continued to fly with his squadrons, leading by example as he had always preferred. He had won his DFC early in the war when he led his 46 Squadron on an attack on several Heinkel 115 float-planes over the North Sea. After Malan left the wing he continued to fly, usually in the Number Two slot to the current Wing Leader. Taking-off for one mission, the engine of his Spitfire had cut out, forcing him to crash land, in which he injured his back. Encased in plaster, he still refused to stay on the ground but continued to fly and fight. Then on the evening of 1 July 1942, he and Squadron Leader Duke-Woolley had scrambled after a lone, high flying German reconnaissance aircraft. Barwell was flying one of the new Spitfire Mark VIs. Unknown to both pilots, two other aircraft had also been scrambled, from RAF Tangmere on the same mission. Barwell's radio must have failed for he did not acknowledge the subsequent radio warning that two other Spitfires were in the area. In the event Barwell was mistakenly shot down into the Channel by one of the Tangmere boys and lost. His place as Station Commander had been taken by Group Captain J.R. Hallings-Pott DFC AFC.

Barwell was not the only high ranking RAF pilot to fall on operations. The famous and much revered Group Captain Victor Beamish DSO DFC AFC was lost in March 1942. It had been

(Right) Biggin Hill early 1943, Sailor with Wing Commander R M Milne, the Wing Leader, and Squadron Leader W A K Igoe, Senior fighter controller of the Biggin Hill Sector. 'Peter' in foreground.

(Below) Sailor talking with Flight Sergeant Vincent Bunting from Jamaica, who flew with 611 Squadron.

Beamish who had discovered the German capital ships *Gneisenau*, *Scharnhorst* and *Prinz Eugen* making their Channel Dash to Norway from Brest the previous month. He had been flying and fighting since June 1940, having joined the RAF in 1923.

The Biggin Hill squadrons had flown against the Channel Dash and on a whole variety of missions throughout 1942. Since Malan had left the scene in August 1941 there was a new and highly dangerous adversary in the air. The Me109s were still around but now the major Luftwaffe fighter was the Focke Wulf 190. They had been superior to the Spitfire Vs but happily, now in 1943, the RAF had the improved Spitfire IX which could match the FW190s in performance.

Since Malan's departure, Wing Commanders had come – and gone. Following Mike Robinson's brief leadership(1), Jamie Rankin had led the wing until December. Then upon his return from America Bob Stanford Tuck took over. When he was brought down at the end of January 1942, Wing Commander Masterman took the leadership. Jamie Rankin returned in April, raising his personal score of victories to 21 by July to receive a bar to his DSO.

Wing Commander E.H. 'Tommy' Thomas DFC, who until then had been CO of 133 Eagle Squadron, led the wing during the summer, including the now famous Dieppe Raid in August. When he left at the end of the year, Wing Commander R.M. 'Dickie' Milne DFC took over, just as Malan arrived. Milne, it will be remembered, had flown with 92 Squadron under Malan and had taken over command of 92 when Malan left Biggin Hill.

*

Upon his arrival at Biggin, Malan found that morale was generally low. Foul weather and a lack of operations was taking its toll. The pilots, indeed, seemed to be spending more time in the cinema than in their Spitfires. With Sailor's arrival things began to change.

The squadrons based at Biggin now were 611 (County of Lancashire) Squadron commanded by Squadron Leader H.T. 'Sinker' Armstrong DFC, and 340 (Ile de France) French

(1)Robinson was lost leading the Tangmere Wing on 10 April 1942.

Squadron, commanded by Capitaine J. Schloesing. At this stage of the war, a fighter wing comprised only two squadrons, not three as in Malan's period in 1941.

Squadron Leader Hugo Throssell Armstrong was 26 years old, hailing from Perth, Western Australia. He had something to live up to, for during the First World War, his uncle, Hugo Throssell, had won the Victoria Cross. At first, Armstrong served without any outstanding distinction with 257 Squadron, then in the summer of 1941, moved to 129 Squadron, part of the Tangmere Wing. He remained with 129 until April 1942, gaining a few combat successes, then he was posted to 72 Squadron as a flight commander. 72 was in the Biggin Hill wing. Soon afterwards he received the DFC, having flown on 29 sweeps and being credited with five victories.

In September 1942 he was given command of 611 Squadron, based at Redhill, Biggin's satellite airfield. By the end of the year he had destroyed at least nine German aircraft and was awarded a bar to his DFC. When Malan arrived 611 had moved from Redhill to Biggin.

It was just as if the Germans were aware that Malan was back, for on 20 January the Luftwaffe paid a visit to the station for the first time in two years. Lunch was being served when the alarm sounded:

'All available aircraft – scramble!' yelled a voice over the Tannoy, 'Bandits approaching from the south-east.'

'Of all the bloody nerve!' was Malan's comment, as he, with others, rushed firstly for the doors and then to the aircraft dispersal areas. Once outside several low flying FW190s could be seen over the northern end of the base. Minutes later a dozen Spitfires were bouncing and roaring into the sky, including Malan, Dickie Milne and Sinker Armstrong.

The raid, by 20 Focke Wulfs, was supported by other diversionary sweeps by more FW190s. The 20 roared over Kent at nought feet but overshot Biggin dropping their bombs on Bromley. One bomb hit a school, killing 45 children and four teachers. The 190s headed out over Croydon before heading for the coast. Malan led the Spitfires due south towards Beachy Head to cut them off and met the German raiders over the coast. In the

(Above) 'Spy' de la Torre, Intelligence Officer Biggin Hill with Sailor in 1943. Squadron Leader H T Armstrong, CO of 611 Squadron, in sweater early 1943.

(Left) Sailor chats to 'Sinker' Armstrong before a mission.

chase and battle which followed, six 190s were shot down, two each falling to Milne and Armstrong, and two to Sous-Lieutenant Robert Gouby of 340 Squadron. As Malan later commented, 'Just like old times.'

*

No sooner had this bit of excitement died down than the boys at Biggin Hill were hit with a tragic loss. It occurred on 5 February. Hugo Armstrong took-off for a routine practice flight over the south coast. In the air was 340 Squadron, having already scrambled to intercept FW190 fighter-bombers which had attacked Hailsham. Armstrong heard the excitement over the radio also hearing them being vectored across the Channel. Armstrong decided to go too.

'Care to join the ladies?' he said to the two pilots with him, who readily agreed. However, the three Spitfires, flying below the clouds, were bounced by eight FW190s of JG26 near Boulogne, who had been escorting the fighter-bombers.

Armstrong's Numbers Two and Three fought for their lives, and heard their leader calmly call: 'This is it chaps. I'm baling out.'

They saw his Spitfire dive towards the sea, belching out black smoke (BS435 FY-F). Blue Two radioed a 'Mayday' call and Dickie Milne took-off with the rest of 611 Squadron to search for the Australian, but they failed to find any trace of him. His victor was Unteroffizier Heinz Gomann of 5/JG26. II Gruppe of JG26 was commanded by Hauptmann Wilhelm-Ferdinand Galland, brother of General Adolf Galland. The Biggin Hill wing often tangled with this *Gruppe*.

Squadron Leader J.H. Slater, a supernumerary squadron leader with 611 Squadron, recorded in his flying log-book:

> Hugo Armstrong bought it – apparently very unlucky, coming through 10/10th cloud was bounced at 300 feet. Called his own Mayday and said he was baling out. Sections never saw a parachute and he must have gone in – a very great loss.

'Garth' Slater had been in the RAF since 1933 and from the start of the war had been retained as a flying instructor in both England and Canada. Having amassed over 2,000 flying hours he finally managed to get onto operations with the Biggin Hill Wing. He was attached to 611 to gain combat experience before a possible move to command a unit of his own.

Armstrong's place was filled by Squadron Leader C. 'Wag' Haw DFM who had gained some local fame when flying with 151 Wing in Russia. He was the holder of the Order of Lenin.

Not long after Armstrong's loss, 340 Squadron too lost its CO. Commandand J. Schloesing, *'le Grand Chleuh'*, was shot down over Le Touquet on 13 February (BS244 GW-P) by a FW190 of II/JG26. His place was taken by Commandant E. Reilhac.

It was at this time that Al Deere again came in close association with Sailor Malan. With the DFC and bar and nearly a score of victories, Deere had just finished a period at the RAF Staff College and then posted to 13 Group Headquarters as a staff officer. However, He was raring to get back onto operations and managed to persuade his AOC, Air Vice Marshal 'Daddy' Bouchier, that any staff officer who had been away from active flying ought to have some up to date idea of current tactics etc: to enable him to carry out his job properly.

As Sailor was OC Biggin Hill, Deere pleaded to Bouchier to have just two weeks on attachment with Malan's wing so as to be brought up to date. (He had been away from operations for six months.) Bouchier was persuaded so Deere hastily went to Biggin; Sailor had already agreed to have him.

Deere was hoping to get his own wing soon, so it was important for him to bag at least one German to show the powers that be that he was the man. Yet bad weather restricted operational flying and after the 14 days had passed, Deere had not fired a single shot in anger. He stayed on for a couple more days, just in case, and then on a mission over St Omer, right at the last, he managed to shoot down a Focke Wulf.

Reluctantly he returned to 13 Group shortly afterwards. Only a few days later he was given command of the Kenley wing. Events, however, were to change this posting.

On 14 March the Biggin Hill wing flew on Rodeo 188, led as usual by Wing Commander Milne. The wing climbed to Dungeness,

Malan with Armstrong and Garth Slater ***(left).*** Armstrong was killed on 5 February 1943, Slater on 14 March.

Wing Commander R M 'Dickie' Milne DSO DFC; prisoner of war 14 March 1943 whilst leading the Biggin Hill Wing.

flying one orbit at 22,000 feet before setting course for Hardelot, gaining more height as they crossed the sea. In mid-Channel condensation trails forced them to orbit again but the trails came from friendly aircraft. Flying again towards Hardelot, Milne led them through cloud.

Soon after crossing into France, Flight Lieutenant F.F. Collerado-Mansfeld, a naturalised American of Austrian extraction, spotted eight German fighters below; Yellow Section of 611 attacked, damaging one of them. Flight Sergeant R.M. McClay (BS510), Yellow Two, (an Australian) did not pull out following the initial attack and was last seen spinning down with a 190 on his tail. Wing Commander Milne ordered everyone who wanted to attack to do so immediately, but his wingman, Flying Officer Garden, developed engine trouble as he started his dive and Milne's Three and Four stayed with Garden until he got his engine going again. Milne was last seen diving to the attack and failed to return.

Number 340 Squadron also attacked but its CO, Commandant Reilhac, and Squadron Leader Garth Slater AFC with 611 Squadron, both failed to return. It was Slater's 22nd operational sortie whilst flying with the wing. He was last seen near Le Touquet attacking an enemy fighter above. The wing retuned in twos and threes to land at Biggin around 6.30 p.m. They had claimed two FW190s destroyed and two damaged. A dear price for the loss of one Wing Leader, one squadron commander, a supernumerary squadron leader and a sergeant pilot. Milne had been flying Spitfire BS240, which he had taken over from the previous Wing Leader, Tommy Thomas. Milne had 15 victories when he was shot down, but happily his capture was later reported by the Germans.

Because of Milne's loss, a new Wing Leader had to be found. Sailor telephoned the AOC and asked for Al Deere. Despite his appointment to the Kenley wing, which he had yet to commence, the request was approved. Thus Al Deere went to lead the Biggin Hill Wing and Johnnie Johnson took over at Kenley.

Johnnie Johnson was to become the RAF's top scoring fighter pilot in Europe, with 38 confirmed victories, the only pilot to top Sailor's score. Yet when he did so in the summer of 1944, he was the first to acknowledge how different had been their respective tasks. Sailor's war had been largely a defensive one against odds, whereas Johnson had always been on the offensive. He never had to contend with the large enemy formations Malan had to battle against, he had time to stalk and flush out the enemy. Johnson says of Malan:

> I never served with him and my contacts were limited to attending his various lectures and meeting him socially on various occasions when he was commanding Biggin Hill. Nevertheless, in the dark days of 1940 he soon became a legendary figure in Fighter Command and was a great inspiration to all we young fighter pilots who first joined fighter squadrons at that time.

Another change at Biggin Hill was that 340 Squadron left, their place being taken by the second Free French fighter squadron 341 'Alsace' Squadron, commanded by Commandant Rene Mouchotts DFC Croix de Guerre. With Malan as Station Commander, the famous Al Deere as Wing Leader and a new French squadron just itching to get into action, all seemed set for an interesting period.

*

Deere had come a long way since the days of Dunkirk. He was an experienced and successful air fighter and was keen to put his own ideas into operation. In this he was totally supported by Sailor who gave him every encouragement to put into effect anything he as Wing Leader thought was right. Deere believed that his squadrons should not only be independent but interdependent. This should also apply to each section but, of course, the whole would still be controlled by him, yet only in so far as routes, timings etc; were concerned, leaving the squadron and section leaders a free hand to act upon their own initiative on sighting enemy aircraft, once having warned the Wing Leader of their intention to attack. Deere's basic idea was a change from a mass controlled attack, to a more profitable form of attack which relied on mutual support between squadrons and sections.

One other item of interest when Deere took over as Wing Leader was that Biggin Hill's total of victories claimed during the war, now totalled 983, just 17 short of the magic figure of 1,000.

Malan in fact referred to this on 1 April 1943, on the occasion of the 25th Anniversary of the formation of the Royal Air Force. On that day all personnel were paraded at Biggin and addressed by its most famous Station Commander. Not unnaturally everyone was gripped by the thought that the 1,000th victory was in sight and a mammoth sweepstake was organised with cash prizes in the offing. After some weeks of collecting, the pilot who shot down the 1,000th victory, stood to win £300, while the person who drew the name of the pilot, would win £150. It certainly was going to be an interesting time in the weeks ahead.

*

The day after the 25th Anniversary Parade, Malan himself led the wing (in JK767) on Circus 279 despite its being his fifth wedding anniversary! He led them away from Biggin Hill at 2.52 p.m.; the wing was flying as Second Fighter Echelon. They crossed the French coast at Berck-sur-Mer, penetrating to St Omer but no enemy aircraft appeared.

The next day 341 and 611 escorted eight bomb-carrying Typhoons (Bombphoons) of 181 Squadron, attacking Abbeville aerodrome. Malan flew as Number Two in Yellow Section to Captain Christian Martell of 341 Squadron. The squadrons orbited at 16,000 feet ten miles off Cayeaux but again no enemy fighters came up. Martell, in Al Deere's opinion, was the best of the French fighter pilots. He often flew as Number Two to Deere; in fact he insisted on flying in that position. He was a big, powerful man, who would often disappear for periods. During these absences he was dropped into occupied France to 'deal' with certain people or problems with the Resistance. Deere recalls that Martell's own mother and father were betrayed to the Germans by another member of the family, who was later dealt with.

Also on the 3rd, Flight Lieutenant E.F.J. Charles DFC, joined 611 Squadron from 64 Squadron. 'Jack' Charles was from Canada, his family having emigrated from England when he was a child.

After a period with the Saskatchewan Horse Militia, he had joined the RCAF in 1938, transferring to the RAF the following year. After a period as a communications, then an army co-operation pilot, he went to 54 Squadron at the end of the Battle of Britain but by then the squadron had moved away from the battle front. He won the DFC in 1941, flying sweeps etc; over France and became a flight commander with 64 Squadron in January 1943. He had 7 victories and at least 7 more probables when he joined 611 Squadron. Already he was becoming known as a coldly efficient fighter pilot. He took command of 611 Squadron shortly afterwards.

Wing Commander Al Deere led the wing on Ramrod 51 on the 4th, with Malan tagging along. They joined the bombers over Beachy Head at 27,000 feet before turning for Le Tréport. On the way over, Deere's Spitfire (JK762) developed engine trouble and had to return; Malan took over the lead. At the French coast they turned north for Abbeville where the Spitfires flew two wide circles but the enemy stayed away. They returned at 2.45 p.m. after 1¾ hours.

Two missions were flown on the 13th. It was a fine day with good visibility. Circus 281 was mounted at 2.10 p.m., Deere was leading, accompanied by Malan. Twelve Venturas of 464 Squadron were to attack the marshalling yards at Abbeville. The wing climbed to 15,000 feet over Biggin before setting course for Hastings on the south coast, crossing at 24,000 feet just after 2.30. The French coast was crossed between Dieppe and St Valéry eight minutes before 3 o'clock, the wing swept round and then out at Cayeaux at 3.11. Yet again the Luftwaffe stayed away.

At 5.15 p.m. 611 Squadron, led by Malan, took-off on Ramrod 56 to St Omer, but still the Germans were conspicuous by their absence. However, on the 18th Jack Charles opened his account with the wing, shooting down a Focke Wulf into the sea off the mouth of the Bay d'Authie.

*

A most unusual event occurred on the day beforehand, or more exactly on the night of 16/17 April. Bill Igoe, by now Senior

Controller at Biggin Hill, records the events in some detail:

'West Malling was at the time the main night fighter station for the Biggin Hill centre, of which I was the Senior Controller. The Sector Commander from Biggin was Sailor Malan. There had been about a week of casual activities, nothing much, and the weather was bad. I was sitting watching the table one night which showed about six raids, three or four marked friendly and the others were marked 'X', three of them were in a triangle area comprising Ipswich, Cambridge and Southend. I had assumed they were Intruders going home and the 'X' probably malfunctioning IFF (Identification Friend or Foe). We had nothing flying because the forecast was bad. To our astonishment, bombs were dropped in the Colchester area whereupon Group turned every raid 'X' and I was asked if I could put some night fighters up to investigate.

'The interesting part of the story, to the best of my memory, I think the squadron commander was (Wing Commander C.M. Miller DFC – 29 Squadron. Author). I told him I did not like the weather but the situation was so unusual I suggested I put a vertical searchlight just south of the Isle of Sheppey and told the pilot to patrol around the searchlight, thereby relieving him of any navigational problems and making it a simple matter of returning to West Malling if the weather clamped down.

'Meanwhile, the raids north of the Thames had turned south but the plotting was bad and the whole situation mystifying. One more lot of bombs was dropped again, roughly in the same area and time went by. I must have had my sole night fighter orbiting in the vicinity of the searchlight for about half an hour when I got a call from the Observer Corps near the searchlight complaining that I had advised him there was one aircraft, a twin, round the light which they had heard clearly to begin with, but now there were two. You can imagine the reaction. We had no plots showing an aircraft approaching the line. Next thing I called up the pilot to tell him there was another aircraft on the landlight, could he see anything? He answered something like "Rubbish, but I'll have a look."

'In the meantime I was checking a list of missing aircraft, single-seaters, and the only one that had been missing, a Defiant, since mid-day could obviously not have been that as petrol would have long since been exhausted. To my astonishment, a little later, the pilot said, "Yes there is something beneath me. I have just seen a shadow twice." Still thinking it was one of our own I said to him, "Right, stay where you are, I shall deflect the searchlight down on West Malling," which I did.

'I have forgotten why, and I think now it was a Saturday night, but it was certainly a night where there was a certain amount of celebration going on because Sailor had gone to London to attend a party. There was a dance that evening at West Malling. We were approaching midnight and of course, a Controller in charge of a Sector was in charge of all operations subject only to Group control who rarely intervened. I advised the duty pilot at West Malling that I thought a strange aircraft was orbiting my light and I would attempt to land him, would he switch on the flare path, which he did. The aircraft came into the circuit and I switched the beam back to the vertical.

'Now for the drama. The aircraft made a good landing and finished up at the end of the runway, taxied back towards the duty pilot's hut from which the duty pilot and some assistants watched the unexpected visitor aircraft, when into the dim light taxied a Focke Wulf 190 and down from which came an astonished German pilot. Almost simultaneously with this, the Observer Corps came back and said, "We have picked up another one." I did not report to Group at this point as I was waiting for the full report from the duty pilot but the arrival of a second aircraft under beam left me little time for reports. I warned the duty pilot that another one was coming in and I deflected the light back on West Malling again advising the patrolling pilot of what was happening.

'West Malling's station commander was advised of what I was doing and he rang up to protest that I was endangering his station. I said I was responsible, not him, and would bring in the second one and was too busy to talk. The duty pilot's quandary was who to obey. I was the Sector Controller but his boss was the station commander. Time was moving on and the second one put down safely, finishing at the end of the runway but the gallant RAF Regiment now being totally awake (there were all sorts of rumours flying about) dashed out in an Armadillo and as they got near the pilot gave a burst of engine to swing his aircraft round to taxi back to the duty pilot, but the Regiment, full of ardour and zeal, not to mention the confusion that existed, thinking the burst of engine indicated an attempt to get away, opened up with a machine gun on the pilot but fortunately were not accurate. The pilot, whose astonishment was even evident to the crew of the Armadillo, got out of the cockpit but was hit by a burst, in the backside. He managed to fall out the other side and was picked up bleeding profusely. However, the drama was not over. The fire brigade turned up because by then some of the bullets had set the aircraft on fire and while attempting to put it down with foam extinguishers it blew up. I cannot remember how many people were killed but the aircraft was a complete wreck.

'I knew I had one completely intact FW190 which was invaluable but now I had a third one on the beam and followed the same procedure to home it into West Malling. In the meantime, a battle was going on. Malling's station commander demanded to speak to Sailor (I fobbed him off temporarily) then to Group Controller, closed the festivities and demanded to speak to Malan, of course, going over my head and I procrastinated saying I was looking for him which was true but I knew it might be hours before I found him.

'Meanwhile, I had to try and land the third one which got on to the circuit and was gliding into land when Malling's station commander ordered the flare-path off. The 190 pilot lost control and spun into an apple orchard. Now all was chaos. I lost number four off the searchlight and some time later when things quietened down I brought the patrolling aircraft into land. Now we had a mess.

'The station commander onto Group, Group onto me, everybody looking for Malan. King's Regulations being thrown round the clock as to who was in control and this was always a ticklish one for the lawyers. In the RAF the pilot of an aircraft is responsible for everything connected with it. The initial argument was, did I control the flare-path or was the station commander responsible for its control in respect of his position as station commander? So the battle raged.

'In the meantime everybody was trying to find Malan. As I did not want to report my side of the story to Group until he returned. I think it must have been four in the morning when he finally turned up into the Ops Room and seemed to have been thoroughly enjoying himself. So we got in a car and drove to West Malling. Dawn was just breaking. With us was the Senior Intelligence Officer, Squadron Leader 'Spy' de la Torre who had been to London with Sailor. Arriving at the duty pilot's hut we found the second German pilot lying on a stretcher, the holes in his bum were very superficial and he was smoking and drinking tea, looking quite happy about the whole business.

'No attempt had been made to find the number three 190, because the Observer Corps had reported that the position of the crash was off the perimeter of the aerodrome. I set out with Malan and found him, an NCO pilot, seated in the living room of a farm cottage, in an armchair in front of a low fire with two elderly people watching him. They had heard the crash and found him entangled in the mess created by the FW190 hitting two apple trees. How they got him out of it and into the kitchen, I don't know. It was the sort of cottage where everything was in one room. The macabre touch about the whole scene was that above the fireplace, smiling benignly down on this poor devil, was the bland cigar smoking face of Winston Churchill on one of those cheap prints which was so common in those days.

'The pilot was fully conscious but his face was quite black and drawn in and he was muttering something in German which we could not understand. He looked a frightful mess. The impact must have been terrific when he hit. The farm labourer was in his late sixties and his wife about the same age. It transpired they had not even spoken to him and had offered him nothing to drink. He had been there for hours by the time we arrived and had no assistance of any description because, as they explained, when they reported the crash and the fact that they carried the pilot into the living room, they had been told they must not on any account offer sustenance or comfort until he was interrogated. We rectified the matter and had him packed off to hospital.

'The outcome of the night's work was that we had now in our possession one completely undamaged Focke Wulf 190, which the

Germans had risked in night intruder operations with bombs. In inspecting the maps we obtained from the undamaged aircraft we found they had been doing casual bombing north of the Thames and on the way back, single-seater navigation being bad at the best of times, they were returning to their base which was about 30 km inside Cap Gris Nez. Their homing beacon (searchlight) quite by coincidence was roughly in the same position the other side of the Channel as the one I had stuck up south of Sheppey. Obviously the pilots had seen the water crossing beneath them which was quite big and that plus the combination of my friendly homing searchlight convinced them they were back a little ahead of their ETA.

'The station commander made a report on the incident, naturally supporting his unilateral action in switching off the flare-path and in justifying what he did by making considerable derogatory remarks on our behaviour. I pass no judgement on the proceedings, you can do that yourself but of course in those days Malan was a considerable figure and the station commander at Malling had been off operational flying for some time. Naturally he was considerably junior to the Sectior Commander that Malan was.'

The captured fighter was an Fw190 A-4. It was test flown against a Spitfire the next day before being shipped to Farnborough. It then went to Collyweston, Northants, where in May 1944 it joined the RAF's Enemy Aircraft Flight – No. 1426 Flight. The unit's CO, Flight Lieutenant E.R. Lewendon was killed when flying this FW190 on 13 October 1944, crashing on the Stamford – Kettering Road, near Collyweston following an in-flight fire.

*

As Wing Leader with a Station Commander of the calibre and fame of Sailor Malan, Al Deere had to a degree quite a problem. Deere remembers that Malan was always keen to go on wing shows but the New Zealander only let him go on certain missions. It was a great responsibility for him and he had to ensure that he always gave Sailor a good Number Two. Malan would often complain bitterly: 'You never let me fly!' to which Deere would reply: 'I'm the Wing Leader and I'll take you when I think it's right.'

Deere knew only too well Malan's past achievements but he was also fully aware that Malan had been away from active operations for a long time. Also, Malan was not getting any younger and one soon lost touch when away from operations for such a long period. Yet it did not stop the South African from flying whenever he could. In the light of evidence of the age of flyers in both the later Korean and Vietnam wars, age apparently had little effect on some air fighters of the later jet age.

Deere led the wing and Malan on escort to Circus 288 on 20 April 1943, 12 Venturas of 487 Squadron attacking the Boulogne marshalling yards. The wing flew from St Pol to St Omer to Calais but again no hostile fighters came up to oppose them.

Twelve Venturas of 21 Squadron attacked the Abbeville marshalling yards on the 21st. It was a good morning but the weather began to deteriorate after lunch. It was Circus 290. Deere led his pilots off at 11.32 a.m.: Malan was flying too in his usual Spitfire JK769. It was just as well, for after making rendezvous with the bombers over Beachy Head, then heading out towards the Somme Estuary, Deere ran into trouble. Nearing the French coast his long range fuel tank would not jettison and nor would that of his Number Three, Flying Officer V.S. 'Paddy' Neill. He had no alternative but to abort and he took his Red Section with him. Malan, therefore, took over the lead of the wing.

Well into France, eight FW190s were seen at 10 o'clock below. Malan sent 611 Squadron down, but the 190 pilots saw the danger and dived away. More Focke Wulfe came into view. Malan swung round with his Yellow Section to dive at them but they also saw the Spitfires too early, nosed away, diving to ground level, and Malan broke off the pursuit at 8,000 feet. Regaining position with the Venturas, the Controller radioed that the bombers wanted 611 Squadron lower and at that moment more 190s attempted to attack the squadron from behind, but all attacks were foiled by the Spitfires turning towards them when they came within firing range. Malan and the wing brought the bombers home safely.

Also on the 21st, Mr Gilbert Harding, together with sound engineers and technicians from the BBC arrived at Biggin for preliminary recordings in anticipation of the 1,000th victory being scored. Excitement was certainly rising.

Another duty Malan had to attend to from time to time, and in April 1943 yet again, was to give an address to various meetings for 'Wings for Victory'. On these do's he often dragged along a protesting Al Deere. Deere hated making these appearances, remembering an experience earlier in the war. On that occasion he had travelled to Glasgow to speak to workers at a gun-cotton factory. He stayed in an hotel in the city and that night the town was bombed. The following day he was mobbed for being in the RAF, for not being in the air defending them, and for not shooting down the raiders. It was not a pleasant experience.

At the end of April came the sad news that Sailor's brother Francis had been killed in action. Francis, a pilot officer, had been flying with 72 Squadron in Tunisia. He had shown a little of his elder brother's flare for action and in his brief operational career was often to the fore. On 28 February 1943 he had flown Spitfire EN311 while escorting Hurricane bombers during an evening raid. Me109s had been engaged, and Malan hit one which was last seen flying towards Bizerta pouring glycol.

Flying as High Cover to Hurribombers on 11 April, against Enfidaville in the early morning, 109s were again met, three being shot down. One fell to Francis Malan (EN301). On the fateful morning of 26 April, Wing Commander H.S.L. 'Cocky' Dundas DFC, OC of 244 Fighter Wing, led his men on a patrol over the Medjez el Bab—Pont du Fahs area; Francis was flying in a Spitfire IX EN294. Flying in two formations of six aircraft between 5,000 and 8,000 feet, enemy aircraft were reported east of Pont du Fahs going south-west. The British formation did a climbing turn and saw flak bursting over Bon Arrada. Six twin-engined Me110s and six Ju88s, escorted by a dozen Me109s and FW190s, were seen coming head-on towards the Spitfires. There followed an inconclusive action. Francis was last seen weaving frantically at about 1,000 feet, chasing a Messerschmitt 110 through heavy flak and did not return.

Francis was not the only Malan, other than Sailor, on active duty. Another brother, Ralph Jordan Malan, was an air observer with the South African Air Force; he eventually rose to the rank of Captain. He had been flying with 24 SAAF Squadron since April 1941, and had seen action in Abyssinia, Madagascar and the Western Desert. During operations culminating in the capture of the island of Madagascar in May 1942, 'Bull' Malan and his crew, had been obliged to force land 300 miles down the eastern coast from Diego Suarez. Under the leadership of their pilot, Major K. Jones, they had begun to walk northwards, taking one of their aircraft's machine-guns along with them. During the trek they saw a group of armed natives under the command of a French officer, coming towards them. Deploying the rest of the crew to cover, Jones met the Frenchman, who informed Jones that he was his prisoner. However, Major Jones then signalled his men who fired a warning burst from the machine-gun, whereupon the Frenchman and his natives became the South African's pirsoners.

Brother Stanley Malan, did not choose the air to fight his war. He was a sergeant in the army, serving in tanks, and was wounded in the Battle for Italy.

Even the 'fighting Malan's' mother, Mrs Evelyn Malan was active despite her age, working in the Pay Corps with the Women's Auxiliary Air Force with the rank of sergeant. A fifth brother, Peter, joined the air force and was trained as a pilot but the war ended before he saw active service.

*

The wing flew as Second Fighter Cover in support of a force of 79 American B17 Flying Fortresses on 4 May; the Americans' target was the Matford Works at Antwerp. The show – officially Ramrod 68 – was engaged by FW190s from JG26. Al Deere destroyed one to bring Biggin's total score to 995. Paddy Neill did not get home.

However, bad weather curtailed further operations for ten days. Then on the 14th, the wing acted as Third Fighter Cover to Ramrod 73, raiding III/JG26s own case at Courtrai–Wevelghem. Again it was American B17s that made the attack, good results being observed. Luftwaffe fighters reacted and in addition to two B17s being shot down, one Spitfire was lost. Its pilot, Sergeant Clarke of 611 Squadron, collided with a Messerschmitt 109G, kill number 996. Jack Charles hammered a Focke Wulf and sent it down to crash and Christian Martell destroyed another 109G. This was 341 Squadron's first official victory but Martel's and Charles' kills brought the Biggin Hill Sector's victory score to 998. It had been a long road since victory number one, scored on 21 November 1939. On that occasion Flying Officer J.A. Davies, an American, and Sergeant Brown of 79 Squadron, had intercepted and shot down a Dornier 17 over the English Channel.

Operations were on for the next day, 15 May 1943. It dawned a fine, bright day. Pilots were called for briefing at 9.30 a.m. The mission was Circus 297, to be flown in three parts. The plan was for six Mitchells of 98 Squadron to bomb the Luftwaffe airfield at Caen-Carpiquet, home of the Richthofen Geschwader – JG2 – and then Typhoon fighter-bombers of 181 Squadron would strike at the same airfield. Four Spitfire Wings supported the operation. Part II of the Circus was for 12 Bostons of 107 Squadron, escorted by five wings, to attack Poix air base, where more than 40 Me109s had been seen by RAF reconnaissance aircraft. Part III called for six Mitchells of 180 Squadron to bomb Abbeville airfield, but this phase was called off.

Such a large operation, as planned, was almost certain to stir up enemy fighters, so Deere suggested to Malan that this was one mission he ought to fly on. The Station Commander gladly accepted his Wing Leader's kind invitation.

Despite the early briefing time, it was not until 4.21 p.m. that the 'tits were pressed'; Deere led 611 Squadron and the wing, with Sailor (JK769) flying as Red Three.

René Mouchotte led his French boys and formed up with 611 Squadron, led by Jack Charles. Everyone seemed to sense that they were flying out to make history. The only questions to be answered was who would actually get the 1000th kill.

The wing roared aggressively into the air and, keeping low, headed south to flash out above Shoreham. At 4.41 p.m. Deere opened the throttle and, followed by his men, began to climb as the French coast came into view. Over their headphones the pilots could now hear the crackle of German radio interference. Then came the voice of the own fighter controller. Bandits were coming up from Amiens, from Rouen and others were circling Lisieux.

Deere levelled out at 21,000 feet crossing into France at

Trouville at 5 o'clock, then he turned the wing towards Caen. As they neared the town of Bonnelosq, the wing fanned out into tactical fours, and then commenced a turn to the right in order to sweep in behind the Caen target area.

Enemy fighters had indeed been stirred up; nine FW190s of I/JG26 attacked the Polish Northolt wing flying as High Cover. They flashed down and through the two squadrons, 315 and 316, just as the Mitchells let go their bombs. The Mitchells went into a steep evasive turn, so steep that several Spitfires were thrown out of position and became momentarily stragglers. This was just what the FW190 pilots were waiting for. They pounced, shooting down Group Captain Stefan Pawlikowski(1) and Sergeant Lewandowski of 315 Squadron.

Al Deere and his pilots, still some 15 miles to the south-east of Caen, saw the distant flak and the 190s and Spitfires in their deathly embraces. Yet it was too far away and the first doubts that today would be 'the day' began to creep into the pilot's minds.

Then quite suddenly, two Focke Wulf 190s were seen below, climbing hard out of a summery haze. I/JG26, based at Triqueville had most of its aircraft unserviceable. However, six or seven 190s of the 2nd Staffel became airborne led by its *Staffelkapitän*, 22-year-old Oberleutnant Horst Hannig. The rest of the I Gruppe attacked the bombers and the Poles. Coming up to battle height, after Hannig's Focke Wulfs had become strung out, the young *Staffelkapitän* and his wingman, Unteroffizier Ernst Godde, slightly alone, were totally unaware that they were climbing directly under one of the RAF's most powerful fighter wings, hidden from them in the glittering sun.

The Wing Leader could see that Jack Charles was in the best position to attack the two FW190s and ordered him and his section down after them. Jack Charles (in EN554 FY-Y) dived to the right as the two 190s headed in towards the bombers. By the time he got within range Charles was approaching them from slightly underneath. Selecting the rearmost 190 he began to fire at 250 yards down to 50 yards, with a four-second burst. Bits came

(1)Pawlikowski had flown in France in 1918 and in the Polish Air Force in the mid-war years. In 1939 he commanded the Polish Pursuit Brigade.

away from the 190's wings and the strikes sparkled along the length of the fuselage on the right side. A fire started beneath the fighter, but then Charles was passing the 190. Crossing over, he closed in behind Hannig's 190, but he started to turn his fighter tightly to the right. The Canadian fired a five-second burst from 180 yards with ¾-deflection. He saw strikes on the wings, followed by an explosion near the cockpit. The 190 slewed to the right, then began to dive. As Charles broke away he saw a parachute open back where the first 190 had been. Flight Lieutenant Johnny Checketts, Yellow Three, saw the pilot of the first 190 bale out and also saw Hannig's 190 dive down and go straight in.

'You've got him, Jack,' yelled Checketts.

'Turban Yellow Leader to Brutus,' radioed Charles to Deere in a matter-of-fact voice, 'both enemy aircraft destroyed.'

Charles had fired 280 rounds of cannon and 1,000 rounds of .303 ammunition in the attack. The time was 5.15 p.m.

In the same moment as Charles was picking off these two 190s, René Mouchotte, at the head of his squadron had been flying west, hoping to bounce any stragglers from the main fight. He spotted a lone FW190 and was after it immediately, giving it a long burst from dead astern. The 190 caught fire and exploded, forcing the Frenchman to turn sharply in order to avoid the shower of debris.

'Good show, Grass-seed leader,' said Captain Michael Boudier, leading Blue Section. Mouchotte radioed his success in the same moment as Jack Charles. So, within seconds of the first burst of cannon and machine-gun fire, all three Focke Wulfe of the 2nd Staffel were falling from the sky. But who had actually scored the 1,000th victory?

It was impossible to say. Charles and Mouchotte had scored their kills at almost exactly the same moment. After debriefing, Malan decided the fairest way was that both squadron commanders should share the honour – and share the prize. And who drew the pilot's name and shared the pilot's sweepstake? Red Three – Group Captain Sailor Malan!

Perhaps it would have been nice if Sailor himself had shot down the thousandth German. Certainly as Fighter Command's top scoring fighter pilot no one else deserved it more. Indeed he

(Above) De-brief! Wing Commander Al Deere listens as Jack Charles describes an air action while Collerado-Mansfeld *(back to camera)* asks questions.

(Left) Squadron Leader Jack Charles celebrates another victory.

had been close. Upon landing he had said to Deere, whilst pointing an accusing finger at the New Zealander.

'You're a fine one, Al, why the hell didn't you let me go after those Huns? They passed right under my nose?'

'And mine too, Sir,' replied Deere, 'but Jack was in the best position. I must say I was tempted to have a go myself – I could do with that £300.'

'Well we'll never know for sure who got the one that mattered,' said Malan. So a Canadian (who was born in England) and a Frenchman shared the 1,000th kill, while a New Zealand Wing Leader and a South African Station Commander looked on.

*

Quite by chance, 341 Squadron had organised a party at the Hyde Park Hotel in London for the night of 15 May. With the day's successes it was obviously going to be quite a thrash. Deere asked Group to have the wing stood down the next day for quite obviously few pilots would be fit to fly. Group operations said that a show had already been planned but promised to use the wing only as a last resort.

The party, as expected, was a tremendous success, but the next morning the inevitable happened. Group required the wing for the show – to be airborne by 10.30 a.m.! At that, for pilots with hangovers, ungodly hour they had to fly down to Portreath to refuel before escorting Venturas on a bomb raid on the Cherbourg airfield at Morlaix. Luckily the Luftwaffe stayed away. A more official celebration for the 1,000th victory was planned, but it did not take place until June.

The following day, the 17th, two days after the great day, representatives from the press descended upon Biggin Hill to cover fully the reporting of this most famous event. As it happened, the wing flew a mission mid-morning, led by Jack Charles in fact. Over the Cabourg area Charles attacked a FW190, and its pilot baled out. This gave an added fillip for the press at Biggin.

A few days later, Al Deere was awarded the DSO – an extremely popular award with everyone on the 'Bump'.

Then came the official celebration of the 1,000th kill, a Ball held on Wednesday, 9 June, held at the Grosvenor House Hotel in London's Park Lane. More than one thousand guests were invited, including Leigh-Mallory, AOC Fighter Command and Arthur 'Bomber' Harris, AOC Bomber Command. A cabaret was provided by Vivian Van Damm and the girls of London's famous Windmill Theatre. For Malan it was a double celebration, for only a few hours before the party began, his wife Lynda gave birth to their second child in the Kent County Hospital. Lynda had wanted a girl to make her family complete and her wish had come true. As Malan said, it was the greatest night of his life.

At the end of the evening there was something of a problem of how to get most of the pilots back to Biggin. Then someone remembered a long-standing promise by a group of London taxi drivers to offer help in just such an emergency. A telephone call brought some fifty cabbies to the rescue. A short time later these same cabbies were invited to be Malan's guests at Biggin as a special 'thank you'.

*

Over the next few weeks, Malan was in the air as much as ever. On 20 May he flew with 611 Squadron on Rodeo 222, as Red Three to Jack Charles. Five days later – a beautiful day according to 611's diary, Malan led the squadron in Circus 304 to Abbeville. Six bombers from 180 Squadron flew the mission, one bomber being hit by flak over the target and exploding. A number of enemy fighters were seen but none stayed when the wing turned to engage them. Then a large force of FW190s were seen coming in from the direction of Le Tréport. Now it was the wing's turn not to play ball, and they headed back to England.

On 30 May Johnny Checketts, another New Zealander, shot down a FW190, 611 Squadron's 100th victory of the war. June began. The pace continued. Malan was beginning to tire again, and little wonder, but he insisted on flying missions between running the station.

Circus 313 on 20 June to Poix led by Deere. On Circus 314 two days later Malan led 611 Squadron to Abbeville-Drucat. It was American Fortresses this time, and one was seen to blow-up in

(Above) Biggin Hill's 1,000th Victory. l. to r.: Bill Igoe, Johnny Checketts, Jack Charles, René Mouchotte, Sailor Malan, Al Deere, (?), Michael Boudier (with pipe).

(Left) René Mouchotte, CO of 341 Free French Squadron and Jack Charles, CO of 611 Squadron share the honour of bringing down Biggin's 1,000th German aircraft, 15 May 1943.

mid-air ten miles north of the target. However, bombs were seen exploding on and around the airfield's dispersal points. Malan flew as Green Three to Deere in his usual Spitfire, adorned as usual with his personal initials AG-M.

Al Deere scored an unusual victory on the 23rd, while leading Green Section of 611 Squadron on Ramrod 100, 2 Group's raid against Meaulte, being a diversion for an American B17 mission. Flying at 11,000 feet just north of Berck, Deere saw two Focke Wulfs coming up behind. The Wing Leader made a sharp turn to the left in order to engage them. The two German fighters, caught out, broke away violently upwards and to the left. However, the number two Focke Wulf went into a high speed stall before spinning equally violently to the right. The German pilot failed to recover from the severity of the spin and corkscrewed into the ground some three miles north of Rue.

René Mouchotte led the wing on the 25th on Ramrod 107. Jack Charles led 611, Sailor flying as his Number Three (EN568). Heavy flak groped for the British aircraft but no enemy fighter appeared. They landed back at Biggin around 3.36 p.m.

*

This was one of the last shows Malan flew on while Station Commander of Fighter Command's premier fighter base. It was sad that his determination to continue flying fighter missions in 1943 was not rewarded by his scoring a victory, but obviously his mere presence was a great encouragement to the younger pilots in the wing.

On 1 July 1943 611 Squadron left Biggin, and their place was taken by 485 New Zealand Squadron. Johnny Checketts took over command of 485, thus remaining at Biggin. The Sector's score, meanwhile, had risen to 1,020.

Tragedy came in August when the great Mouchotte failed to return from one of the first missions against the newly located V1 rocket launch sites. Al Deere's Spitfire was unserviceable so he borrowed one from Christian Martell's Flight. Fate, however, decreed that Deere would not fly this mission that afternoon. As he took-off his engine failed and he had to brake to a halt.

Mouchotte, who had led the wing on several occasions of late, took over lead position.

It was a stifling hot day. The Germans reacted to the mission, and Focke Wulf 190s were reported up from Abbeville, St Omer and Hardelot. It was going to be warm in more ways than just the weather. In fact the Luftwaffe really got stuck into the raiders and Mouchotte and his Number Two, Pierre Clostermann, became separated. Clostermann(1) shot down a Focke Wulf but then over the R/T he heard Mouchotte calling, 'I am alone.' They were his last words.

Ten days later Johnny Checketts went down during a battle with FW190s. He shot down one but when out of ammunition he and his wingman were assailed by eight 190s and he was hit. The loss of both Biggin Hill squadron commanders in so short a period was a severe blow to the wing but 341 was taken over by Bernard Duperior, 485 by Marty Hume. In the event, Checketts survived. Although burned on the face and hands he baled out successfully, evaded capture and with help from loyal Frenchmen he was back in England in a matter of weeks. He went on to become a Wing Leader.

*

One of Sailor Malan's last functions as OC Biggin Hill was the unveiling of a memorial to pilots of the Biggin Sector who had given their lives in four years of war. This event occurred on 19 September 1943. The memorial was dedicated at a service in St George's Church in Biggin Hill Camp, by the Chaplain in Chief of the RAF; the lesson was read by Marshal of the RAF, Viscount (Hugh) Trenchard. Among those in attendance were representatives of French, Polish, Australian, New Zealand, South African and Canadian Air Forces as well as the Royal Air Force.

The following week Al Deere left Biggin Hill. He had flown on more than 120 operations as Wing Leader, having raised his total combat victories to 22. He had become tired and was suffering from stomach pains which developed into an acute enteritis. Air Vice-Marshal Hugh Saunders, AOC 11 Group, conferred with

(1)Clostermann's book *The Big Show* is a classic about his war flying.

Malan, deciding that Deere had done more than enough and should be rested. After a period in hospital Deere went to command the Fighter Wing of the Central Gunnery School and later as a staff officer at 11 Group. Bernard Duperior took over command of the wing at Biggin.

Deere was not the only one tiring. Sailor too needed a rest from both operations and running a busy fighter station. On 16 October he was taken ill at his home, and Squadron Leader K.W. Marten took over temporary command of Biggin. On the 7th, Sailor took sick leave; his time at Biggin was finally over. Group Captain H.L. Maxwell DSO took command, and Malan was posted to command Number 19 Fighter Wing with effect from 1 November.

*

Over the next months, Sailor helped to advise and train the pilots in his wing during the long build up to D-Day – the invasion of Northern France. By the spring of 1944, North Weald became 132 Airfield within 2nd Tactical Air Force, which with 134 Airfield, comprised 19 Fighter Wing. Malan's Wing Leader was Wing Commander F.D.S. Scott-Malden DSO DFC.

Then in March Malan took command of 145 Wing, whose Wing Leader was another experienced air fighter, Wing Commander Roy Marples DFC. However, Marples was killed in a collision with a French pilot of 329 Squadron on 26 April. Wing Commander W.V. Crawford-Compton DSO DFC took over on the 30th.

In 145 Wing, Malan was back with old friends. Its three squadrons were all French units, 340, 341 and 329 Squadrons, all flying Spitfire IXBs. Commandant J.A.M. Fournier commanded 340, Christian Martell 341 and Commandant P.C. de G. Fleurquin, 329 Squadron. The wing was based at Merston as part of 84 Group, 2nd TAF.

It was a great time for flag-waving, and visits from various VIPs were frequent. On 17 May 1944 Merston was visited by Generals Koenig and Valin, the latter was the Chief of Staff of the Free French Air Forces in Britain. Malan accompanied them during the visit in company with Air Commodore Beaumont and Lieutenant-

Sailor Malan, Station Commander Biggin Hill talking to Jack Charles and his Wing Commander Flying, Al Deere.

Preparing for D-Day, 1944. Sailor discusses tactics at Merston just prior to the invasion. l. to r.: Lieutenant Raoul Duval, Sailor, (?), Christian Martell and Wing Commander W V Crawford-Compton. Duval had been shot down in April 1943, near his home town in France but escaped back to England, bringing a bride with him.

Colonel Bernard Duperior. The Generals inspected 329 and 341 Squadrons who were drawn up at 340's dispersal for the occasion. D-Day was now just twenty days away.

When the great and momentous day finally arrived there was great excitement and not only at Merston. Wing Commander Al Deere was now given command of 145 Wing which he joined early in June. On the 4th all the senior 2nd TAF officers concerned with the invasion attended a conference at Uxbridge. To everyone it was obvious that the big day was near, although the actual date was still secret. Indeed, even at this stage the date was not certain even to the planners. But the following day the go-ahead was given. At Merston, Deere briefed all the pilots at 10 p.m. in the evening – it was great news!

The following day, D-Day, 6 June 1944, everyone was in the air. And in the late afternoon Sailor Malan decided that he was not going to miss being a part of it. The wing had already flown three sorties during the day. In his Spitfire, adorned still with his AG-M, he led Yellow Section of 340 Squadron on Ramrod 976. His section consisted of Sergeant De Raynal, Lieutenant Porchon and Sergeant C. Rosa. The Spitfires took-off at 8 p.m., 329 and 340 Squadrons escorting RAF Albemarles towing Horsa gliders into the assault area. They made rendezvous over Bognor Regis and then flew out towards France.

Flak over the drop zone was severe, one glider being shot down, but the other gliders went in and the paratroops could be seen landing on either side of the Orne River. An R-boat was seen on the Caen Canal and shot up and damaged by 329 Squadron. The Spitfires landed back at around 10 p.m.

Sailor Malan had come a long way, but D-Day was the culmination of his war. He had flown in action on the day the Allies went back to France. It was not the end, but it was the beginning of the end.

CHAPTER ELEVEN

Finale

In July 1944, Sailor Malan became Commanding Officer of the Advanced Gunnery School at RAF Catfoss. He left 145 Wing just as his old 74 Tiger Squadron joined the two French units, 340 and 341. So right at the end of Malan's active war, he had still been in close association with his famous Tigers.

At Catfoss expert air fighters of all nationalities, came to pool their ideas and develop new techniques and tactics. Under his own experienced fighters eyes he had such men as Don Kingaby DSO DFM, Stanislaw Skalski DSO DFC the Polish ace, Pierre Clostermann DFC the Frenchman who had been with the wing at Biggin in 1943. Jack Charles DSO DFC of the 1,000th victory fame, Eric 'Timber' Woods DFC and George 'Screwball' Buerling DSO DFC DFM, both of whom had gained fame over the island fortress of Malta. There were also a number of Americans including Dick Bong with 37 victories over Japanese aircraft in the Pacific war. It was certainly a vast amount of experience and expertise gathered together.

Malan remained at Catfoss until the New Year. It was then rumoured that he was going to Rhodesia to help train pilots but this proved untrue. Then suddenly the war was over. Sailor Malan had done his bit and more, towards helping to achieve the final victory. The Malan family too had served well and had counted the cost. Francis had died in Tunisia and Ralph too had given his life. Still with 24 SAAF Squadron in Italy and raiding northwards towards Germany he had been killed on 12 June 1944, six days after D-Day in Europe. Only Sailor and Stanley survived.

Soon after the cheering and celebrations at peace being won ended, the question arose as to what Sailor would do next. He was nearing his 35th birthday so it seemed fairly certain that this, together with his recent posts, it would be a desk job for him, or perhaps again the command of an air base. And who could tell,

Group Captain A. G. Malan, D.S.O., D.F.C.,
and the Officers of the Biggin Hill Sector
request the pleasure of the Company of

S/L W. Igoe and Lady

to a Dance at Grosvenor House, Park Lane, London, W. 1.,
on Wednesday, the 9th of June, 1943,
to commemorate the shooting down of the 1,000th Hun Aircraft
by Pilots of the Sector.

9 p.m. - 3 a.m.

This invitation must be produced to gain admission.

Biggin Hill - Gravesend - West Malling - Hawkinge - Lympne.

448

(Above) Bill Igoe's invitation to celebrate Biggin Hill's 1,000th victory.

(Right) Sailor and Wing Commander R A B Learoyd VC at a charity social function.

possibly the chance of Air rank. In the meantime, he attended a six-month course at the RAF Staff College with a view to preparation for a future post in the Service. However, in the final analysis, Malan decided against continuing his career in the Royal Air Force. The excitement of flying, the possibility of action, all had passed with the flush of exuberant youth. With the war over and with a growing family he decided to return to his native South Africa. There he could have a permanent home for his wife, son Jonathan and daughter Valerie – put down roots that even life in the RAF had not really given him, roots he had been seeking since his sailing days. Since his marriage to Lynda he had had nearly thirty different places loosely called home. It was time for a change.

Through friends in the Diamond Trading Company in England, he met Sir Ernest Oppenheimer and his son Harry. Harry Oppenheimer had fought in the Desert with the 4th Armoured Car Regiment in the war. He had been educated at Charterhouse, Oxford, and was heir to the vast empire of some forty companies in the gold-mining, diamond mining and chemical industries in South Africa. The largest of these was the Anglo-American Corporation, based in Johannesberg. Sailor and Harry were of about the same age and discovered they had much in common. After some discussion Sailor accepted an invitation to join the staff of the Johannesberg company.

Before leaving he attended a last farewell party at the White Hart Inn at Brasted in Kent. This was the wartime haunt of the pilots who flew from RAF Biggin Hill and where the famous black-out screen was signed in chalk by most of the great fighter pilots who had served at Biggin. On this occasion the screen, suitably protected by glass and framed, was formally unveiled in the bar parlour. About him were many of his old friends, fellow air-fighters who had survived the war. As he, the greatest of them all, unveiled the screen, he said:

'My feelings are very mixed. This is a happy occasion in some ways, and I am trying to smile. But it would be a strange man who didn't have sadder thoughts when he reads some of these names and remembers. We built up a "Fighter" tradition that will not easily die. I hope you will all come here often to steep yourselves in a bit of tradition.'

*

Malan and his family travelled to Cape Town, South Africa aboard the *Caernarvon Castle*, passage free – a gift from the Union Government. Once he had settled down he was made private and political secretary to Harry Oppenheimer. The younger Oppenheimer was all for the ideals which he, Sailor, and many other South African ex-servicemen believed in, that they had all fought for. Like Sailor he had a strong sense of loyalty to South Africa, despite the fact that Sailor had been away from his native homeland for such a long time. The political scene was never far away from their lives, for Sir Ernest had held the Kimberley seat for the old South African Party and later the Union Party from 1924–1928.

Meanwhile, Malan joined various golf clubs, shortly reducing his golfing handicap to 12, and was generally feted by everyone he met. The hero was home at last. At work he was behind a desk but he found it all very fascinating, yet such a far cry from the cockpit of a Spitfire.

In 1948 the Nationalists Party won power in the South African elections. The party was anti-English in many respects and anti-English in South Africa. Many changes came about and many not for the better. In his capacity as political secretary to Oppenheimer, who was himself a strong Parliamentarian, Sailor was very close to all the political happenings in his country and did not like much of what he saw happening. In an attempt to get away from it all he spoke to Oppenheimer about his future and it was agreed that Malan should leave the Company, wishing to return to the land.

It was many years since he had been close to nature and had first to relearn many of the arts – the craft of farming. He went to Oppenheimer's farm at Mauritzfontein in the Kimberley region. It was a racing stable mainly but he relearnt many things about animals and farming in general.

Late in 1950 Malan took out a mortgage on a farm of his own,

Benfontein of 27,000 acres, which was situated eight miles outside Kimberley. He also purchased 900 sheep. For some months Sailor Malan became the country gentleman. He and Lynda replanned the dilapidated farmhouse while his children were schooled in Kimberley.

However, the political scene was never far away. In 1951 the Nationalists put forward a Bill which would defranchise the Cape Coloured voters, which, if passed, would be against the terms of the Act of Union of 1910, which protected the rights of the Coloured South Africans. This could not be ignored by the Parliamentarians and from this force of will against the Separate Representation of Coloured Voters Act, as the Bill was later called, began the Torch Commandos. Not that it was particularly the rights of certain peoples being infringed that upset so many people as the intention to ignore the Act of Union constitution.

Voices began to be raised in many parts of South Africa. Ex-servicemen banded together and in Kimberley too people became angry. Sailor, who was Kimberley's own local hero, and no stranger to anyone, was caught up in the turmoil. He was a popular figure and anything he cared to say would be taken as being fair and honest. He was asked to attend a rally outside the City Hall in Johannesberg and he spoke to a huge crowd about freedom, freedom just as important as that just fought for in the recent war. It was stirring stuff.

There is no intention in this book to go deeply into the politics of South Africa. Malan, the air hero, became willingly or not, a figurehead, a rallying point for the Torch Commandos. He became National President of the Commandos in 1952, and those who wish to know more of this whole scene are referred to the final Chapter in Oliver Walker's earlier biography of Malan, *Sailor Malan* published by Cassell & Co Ltd, in 1953. This has been a book about Malan the airman, Malan the air fighter, the leader. In peace he remained a farmer, never happier than when he was on his own land with his family.

*

As the 1950's drew on it became evident that Sailor Malan was not in the best of health. The years of strain had taken their toll. He saw doctors in South Africa and also underwent hospital treatment in the Union. He was suffering the first signs of Parkinson's disease which attacks the central nervous system. It can affect people in different ways. In some it affects just the body, in others the mind; in others it affects both.

In Sailor's case it attacked mainly the body. He began to be unsteady on his feet and also to lose weight, yet he still retained his rugged good looks as he approached his fiftieth birthday.

Then in mid-1959 Malan returned to England for the first time in thirteen years. He flew from South Africa to London Airport, and told a reporter:

'I haven't really flown since 1946 but on this trip I took the controls over the Belgian Congo. Couldn't do much with it though. These airliners have to keep on route.'

Meeting him at the airport were some of his old pals from the war years, led by Al Deere and his wife Joan. Also there to greet him were four young Hunter pilots from the present day 74 Squadron who had come down from Coltishall Norfolk with an invitation for Malan to visit the new Tigers. In a brief interview with the newsmen he told them:

'I have Parkinson's disease. Yes I am ill and I plan to see Sir Russell Braine, the specialist, while I'm here, but it's not all that serious. I feel all right. And anyway, that's not my main reason for this trip. The other and more important reason is to see a lot of old friends. And I want to 'do' London and its shows, theatres and to catch up with old memories.'

The date of his visit was 7 June, and he stayed in SW1 while he visited his old friends and old haunts. Then came the visit to RAF Coltishall to meet 74 Squadron. Sailor and Lynda, on this visit, were the guests of Coltishall's Station Commander, Group Captain Harold Bird-Wilson DSO DFC AFC, who like Malan was a former Battle of Britain pilot, having flown with 17 Squadron. The Malan's stayed at 'Birdie's' house and during the two-day stay he

Sailor with Squadron Lead
P W Carr, the CO of 7
Squadron in 1959 during th
visit to Coltishall.

Malan with a 74 Squadro
Hunter with the Tig
markings.

With Squadron Leader Ca
Malan inspects the cockpit
a Hawker Hunter.

took them to the Ferry Inn at Horning, a hostelry frequented by pilots from Coltishall, especially during the war. As will be remembered, 74 had been based at Coltishall for a period in 1940. On the way to Horning they passed the house where Sailor and Lynda had lived during that period. Going into the Inn, Bird-Wilson was in uniform but Malan, naturally, was in civilian attire. No sooner had they entered than one of the locals saw him, stood up, came over and shook Sailor's hand, saying, 'You're Sailor Malan!' It made Sailor's day.

During the short visit to 74, now flying Hawker Hunter F6 fighters, they got Sailor seated in the cockpit of a Hunter which he later signed under the edge of the cockpit, the signature being varnished over later to preserve it. As he sat there at a specially pre-arranged moment, there came a sound from the past. Malan's body might be failing him but his mind and his hearing were still sharp. He looked up towards the sound, the old familiar, beautiful sound of a healthy Merlin engine. His friends and colleagues had arranged for a Spitfire to fly over Coltishall. It proved a magic moment. The years fell away as Malan's keen eyes clung to the sight of the Spitfire as it wheeled and turned above the aerodrome. Then it landed and taxied up to the Hunter. Sailor climbed down from the jet and shortly afterwards was sitting again in the cockpit of the aeroplane in which he had made his name in history – the aeroplane in which he had become Britain's premier fighter pilot.

After the visit, Sailor wrote to the Commanding Officer of 74, Squadron Leader P.W. Carr:

> I am sorry that we had so short a time together at Coltishall. I was so delighted to have seen the squadron although it was for such a short time.
>
> I envy you taking over such a fine squadron. I joined it straight from FTS, commanded a flight within three months and was a flight commander for three years and squadron commander for about nine months – in fact the only squadron in which I served. The spirit was always absolutely first class both pre-war and during the battle. I think I am correct in saying that when I handed it over to Mungo-Park we had either 156 or 176 confirmed kills with the loss of only

Sailor!

Sailor back in the cockpit of a Spitfire after 13 years, Coltishall, 1959.

12 pilots. Practically all the kills were enemy fighters. We also destroyed two bombers at night during the first night raid of the war – ask the STO. I shot down both in one sortie.

Judging by what I saw the other day the squadron is just the same as it always was. One can see that the chaps are proud of the fine record in two world wars.

One doesn't want to have it proved, but if they had to meet the crisis they would acquit themselves as in the past.'

Four years later he was dead. In those final years he had steadily become a chronic invalid. So died Sailor Malan – the greatest of the Few. The date was 17 September 1963. His wife and daughter were with him. He was fifty-two.

*

A month later a memorial service was held in the tiny chapel at Biggin Hill, attended by many of Malan's former friends and contemporaries. More than eighty people were there. Among them was Lord Dowding, who came to pay his respects and his own tribute to his South African ace pilot. Within scrambling distance of the runway from where Malan took-off so often, his friends thought of him as he had been, each said their own farewell.

The address was given by the Reverend Cecil King, station chaplain, who said:

'He belonged to that happy band of brothers round whom our future lay in 1940. He had an invincible spirit to win at a time when we needed it.'

The then Chief of the Air Staff, Air Chief Marshal Sir Charles Elworthy, said of Malan:

'We mourn the passing of a fine officer and gallant comrade who was in the forefront of the Battle of Britain, a man who fought bravely and led fearlessly at a time when courage and leadership in the air was our only safeguard. His name will live on in the history of the Royal Air Force and in the minds and hearts of men of courage everywhere.'

Malan once said: 'Taking a Spitfire into the sky in September 1940 was like entering a dark room with a madman waving a knife behind your back.' Yet he continued to enter that dark room constantly and survived. He will always be remembered as the RAF's greatest air fighter.

Appendices:

APPENDIX A

Record of Service

Short Service Commission as Acting Pilot Officer on probation, General Duties Branch RAF – Civil Flying School, Filton	–	6 January 1936
RAF Depot Uxbridge	–	
3 Flying Training School Grantham	–	
74 Fighter Squadron, Hornchurch	–	20 December 1936
Acting Flight Lieutenant	–	2 March 1939
Flight Lieutenant	–	6 July 1940
Squadron Leader	–	8 August 1940
Wing Leader RAF Biggin Hill	–	10 March 1941
CFI 58 OTU	–	August 1941
USA trip	–	October 1941
CO Central Gunnery School, Sutton Bridge	–	December 1941
Group Captain	–	October 1942
Station Commander RAF Biggin Hill	–	1 January 1943
CO 19 Fighter Wing	–	1 November 1943
CO 145 Fighter Wing, 2 TAF	–	March 1944
CO Advanced Gunnery School, Catfoss	–	July 1944
RAF Staff College	–	1945-46

APPENDIX B

Group Captain A.G. Malan's known combat victories 1940-41

Date	*Type*	*Result*	*Time*	*Duty*	*Locality*	*Remarks*
21 May	Heinkel 111	Probable	7.20-40pm	Patrol	Nr Dunkirk	
21 May	Junkers 88	Destroyed	7.20-40pm	Patrol	Nr Dunkirk	
21 May	Junkers 88	Damaged	7.20-40pm	Patrol	Nr Dunkirk	
22 May	Junkers 88	Destroyed	5.45am	Patrol	N of Calais	1/3 Shared
24 May	Dornier 17	Destroyed	11.35am	Patrol	E of Calais	Shared
24 May	Heinkel 111	Destroyed	3.40pm	Patrol	E of Dunkirk	
27 May	Messerschmitt 109E	Destroyed	9.05am	Patrol	Nr Dunkirk	
27 May	Dornier 17	Damaged	9.30am	Patrol	NW of Calais	
27 May	Dornier 17	Probable	4.00pm	Patrol	St Omer	½ Shared
19 Jun	Heinkel 111	Destroyed	0.20am	Night P	Foulness	
19 Jun	Heinkel 111	Destroyed	0.40am	Night P	Chelmsford	
12 Jul	Heinkel 111	Destroyed	4.30pm	Patrol	Off Margate	1/3 Shared
19 Jul	Messerschmitt 109E	Probable	3.55pm	Patrol	Dover	
25 Jul	Messerschmitt 109E	Damaged	am	Patrol	Channel	
28 Jul	Messerschmitt 109E	Destroyed	2.00pm	Patrol	Dover	
28 Jul	Messerschmitt 109E	Damaged	2.00pm	Patrol	Dover	
11 Aug	Messerschmitt 109E	Destroyed	8.00am	Scramble	Channel	
11 Aug	Messerschmitt 109E	Damaged	10.30am	Scramble	Channel	
11 Aug	Messerschmitt 109E	Destroyed	2.00pm	Patrol	Off Margate	
13 Aug	Dornier 17	Destroyed	7.00am	Scramble	Estuary	
13 Aug	Dornier 17	Probable	7.00am	Scramble	Estuary	
11 Sep	Junkers 88	Damaged	4.30pm	Patrol	London	
11 Sep	Junkers 88	Destroyed	4.30pm	Patrol	Biggin Hill	
17 Oct	Messerschmitt 109E	Probable	3.30pm	Scramble	N of Ashford	
22 Oct	Messerschmitt 109E	Destroyed	2.10pm	Patrol	S of Maidstone	
23 Nov	Messerschmitt 109E	Destroyed	1.00pm	Patrol	Channel	
27 Nov	Messerschmitt 109E	Destroyed		Patrol	Channel	
27 Nov	Messerschmitt 109E	Destroyed		Patrol	Channel	½ Shared
2 Dec	Messerschmitt 109E	Destroyed	12.10pm	Patrol	Off Dungeness	

Date	*Type*	*Result*	*Time*	*Duty*	*Locality*	*Remarks*
1941						
2 Feb	Messerschmitt 109E	Destroyed	2.00pm	Circus 2	Boulogne	
5 Feb	Dornier 215	Destroyed	1.50pm	Patrol	S of Dover	Shared
17 May	Messerschmitt 109F	Probable	12.50pm	Patrol	E of Dover	
21 May	Messerschmitt 109F	Damaged	6.00pm	Circus 10	Channel	
17 Jun	Messerschmitt 109F	Destroyed	7.35pm	Circus 13	NW of Bethune	
21 Jun	Messerschmitt 109F	Destroyed	4.35pm	Circus 17	Boulogne	
21 Jun	Messerschmitt 109F	Destroyed	4.35pm	Circus 17	Le Touquet	
22 Jun	Messerschmitt 109F	Destroyed	4.15pm	Circus 18	Dunkirk	
23 Jun	Messerschmitt 109F	Destroyed	8.25pm	Circus 20	SE of Boulogne	
23 Jun	Messerschmitt 109F	Destroyed	8.25pm	Circus 20	SE of Boulogne	
24 Jun	Messerschmitt 109F	Destroyed	8.40pm	Circus 21	Off Gravelines	
25 Jun	Messerschmitt 109F	Destroyed	12.30pm	Circus 22	St Omer	
26 Jun	Messerschmitt 109	Probable	am	Circus 24	Gravelines	
28 Jun	Messerschmitt 109E	Destroyed	9.10am	Circus 26	Gravelines	
30 Jun	Messerschmitt 109E	Destroyed	6.30pm	Circus 27	NW of Lens	
2 Jul	Messerschmitt 109E	Destroyed	1.00pm	Circus 29	Lille	½ Shared
3 Jul	Messerschmitt 109E	Damaged	11.40am	Circus 30	St Omer	
3 Jul	Messerschmitt 109E	Damaged	11.30am	Circus 30	St Omer	
4 Jul	Messerschmitt 109E	Damaged	3.30pm	Circus 32	St Omer	
4 Jul	Messerschmitt 109E	Destroyed	3.30pm	Circus 32	St Omer	½ Shared
4 Jul	Messerschmitt 109	Destroyed	3.30pm	Circus 32	St Omer	
4 Jul	Messerschmitt 109E	Damaged	3.30pm	Circus 32	St Omer	
5 Jul	Messerschmitt 109E	Damaged	1.14pm	Circus 33	Nr Lille	
6 Jul	Messerschmitt 109F	Destroyed	pm	Circus 35	Nr Gravelines	
23 Jul	Messerschmitt 109F	Damaged	pm	Circus 60	St Omer	
24 Jul	Messerschmitt 109	Damaged	2.01pm	Circus 61	Hazebrouck	

Total listed as 34 destroyed or shared destroyed, 7 probably destroyed and 14 damaged. The 35th victory cannot be readily identified.

APPENDIX C

Sailor Malan's Decorations

Award promulgated in the *London Gazette*, 11 June 1940:

Distinguished Flying Cross

Flight Lieutenant Adolph Gysbert Malan (37604), Royal Air Force.

'During May, 1940, this officer has led his flight, and on certain occasions his squadron, on ten offensive patrols in Northern France. He has personally shot down two enemy aircraft, and probably, three others. Flight Lieutenant Malan has displayed great skill, courage and relentless determination in his attacks upon the enemy.'

*

Award promulgated in the *London Gazette*, 13 August 1940:

Bar to Distinguished Flying Cross

Flight Lieutenant Adolph Gysbert Malan (37604), Royal Air Force.

'Since the end of May, 1940, this officer has continued to lead his flight, and on many occasions the squadron, in numerous successful engagements against the enemy. During the Dunkirk operations he shot down three enemy aircraft and assisted in destroying a further three. In June, 1940, during a night attack by enemy aircraft, he shot down two Heinkel 111s. His magnificent leadership, skill and courage have been largely responsible for the many successes obtained by his squadron.'

*

Award promulgated in the *London Gazette*, 24 December 1940:

Distinguished Service Order

Acting Squadron Leader Adolph Gysbert Malan DFC, (37604),

Royal Air Force No. 74 Squadron.
'This officer has commanded his squadron with outstanding success over an intensive period of air operations and, by his brilliant leadership, skill and determination has contributed largely to the success obtained. Since early in August 1940, the squadron has destroyed at least 84 enemy aircraft and damaged many more. Squadron Leader Malan has himself destroyed at least eighteen hostile aircraft and possibly another six.'

*

Award promulgated in the *London Gazette*, 22 July 1941:
Bar to Distinguished Service Order
Acting Wing Commander Adolph Gysbert Malan DSO DFC (37604), Royal Air Force:
'This officer has displayed the greatest courage and disdain of the enemy whilst leading his Wing on numerous recent operations over Northern France. His cool judgement, exceptional determination and ability have enabled him to increase his confirmed victories over enemy aircraft from 19 to 28, in addition a further 20 damaged and probably destroyed. His record and behaviour have earned for him the greatest admiration and devotion of his comrades in the Wing. During the past fortnight the Wing has scored heavily against the enemy with 42 hostile aircraft destroyed, a further 15 probably destroyed and 11 damaged.'

*

Foreign Decorations.
In addition to the above, Sailor Malan was awarded the following decorations by Allied Governments:

Belgian Croix de Guerre
Czechoslovakian Military Cross
French Legion of Honour, in the degree of Officer
French Croix de Guerre, with Palm

APPENDIX D

The Malan Memorial Sword

The Malan Memorial Sword was presented to No 74 (Fighter) Squadron RAF on 15 July, 1966 by a number of former members of the Squadron, of the years 1936 to 1945, in proud memory of Sailor Malan, who served with the Squadron in all ranks from Acting Pilot Officer to Squadron Leader, Commanding. Sailor was continuously associated with the Squadron for over five years and served with no other Squadron, a record of service rare in itself.

It is intended that the Sword should serve as an inspiration to those coming after, so that Malan's high standard of courage, determination and leadership shall live on in honour of his exceptional service to the Squadron.

The donors, hoped that the Sword would be carried by the Squadron Commander on ceremonial occasions, and permanently displayed on the Squadron. The following former members of the Squadron were associated with the presentation of the Sword:

Group Captain H.D.P. Bisley CBE AFC
Group Captain D.S. Brookes*
Wing Commander V.G. Byrne
Air Commodore H.G. Crowe CBE MC*
Flight Lieutenant A.R. Griffin
Group Captain D.S. Hoare
Squadron Leader P.F. Illingworth*
Flight Lieutenant J.R. Lewis
Squadron Leader E. Mayne AFC
Squadron Leader W.W. Page
Squadron Leader K.J. Plested AFC AFM
Wing Commander T. Rowland
Wing Commander H.M. Stephen DSO DFC
Squadron Leader D.P. Tidy MA
Wing Commander O.H.D. Blomfield
Wing Commander G.E. Burtenshaw
Group Captain E. Carr
Squadron Leader T.R.W. Froud
Squadron Leader J.C.F. Hayter DFC*
Group Captain H.B. Hurley
Group Captain D.P.D.G. Kelly DFC
Group Captain P.G.H. Matthews DFC*
Squadron Leader E. Moody
Squadron Leader F.G. de Pass
Wing Commander P.H.M. Richey DFC*
Flight Lieutenant W.M. Skinner DFM
Squadron Leader H. Szczesny VM DFC KW
Group Captain F.L. White*

*former Commanding Officers of 74 Squadron

In 1971 the Sword was presented to HQ 11 Group, Bentley Priory for safe-keeping, being received by the AOC Air Vice Marshal I.G. Broom.

BIBLIOGRAPHY

Sailor Malan, Oliver Walker – Cassell & Co Ltd, 1953
Tiger Squadron, Ira Jones – W.H. Allen, 1954
I Fear No Man, Douglas Tidy, Macdonald & Co, 1972
Nine Lives, Al Deere – Hodder & Stoughton, 1959
The Mouchotte Diaries, Ed. Andre Dezarrois – Staples Press Ltd, 1956
2 Group RAF, M.J.F. Bowyer – Faber & Faber 1974
Battle over Britain, F.K. Mason – McWhirter Twins 1969
Fighting in the Air – Arms & Armour Press 1978
Under the White Rose, Frank Ziegler – Macdonald & Co, 1971

INDEX – NAMES

A separate index to RAF Units and Bases follows on page 219

SQUADRONS

RAF BASES